SCALING AND PARTICULATE FOULING IN MEMBRANE FILTRATION SYSTEMS

"Whoever undertakes to set himself up as a judge in the field of truth and knowledge is shipwrecked by the laughter of the gods."

Albert Einstein

Scaling and Particulate Fouling in Membrane Filtration Systems

DISSERTATION
Submitted in fulfilment of the requirements of
the Academic Board of Wageningen University
and the
Academic Board of the International Institute for Infrastructural,
Hydraulic and Environmental Engineering
for the Degree of DOCTOR
to be defended in public
on Tuesday, 19 June 2001 at 11:00 h
in Delft, The Netherlands

by

SIOBHAN FRANCESCA E. BOERLAGE

born in Auckland, New Zealand

CRC Press
Taylor & Francis Group
Boca Raton London New York

CRC Press is an imprint of the
Taylor & Francis Group, an **informa** business

This dissertation has been approved by the promoter:
Prof. dr. ir. J Schippers, IHE Delft/Wageningen University, The Netherlands

Members of the Awarding Committee:
Prof. dr. ing. R. Gimbel, University of Duisburg, Germany
Prof. dr. W. Rulkens, Wageningen University, The Netherlands
Prof. dr. G.J. Witkamp, Delft University of Technology, The Netherlands
Prof. ir. J.C. van Dijk, Delft University of Technology, The Netherlands
Dr. M.D. Kennedy, IHE Delft, The Netherlands
Prof. dr. ir. G.J.F.R. Alaerts, IHE Delft/TU Delft, The Netherlands

Contents

Acknowledgement

Firstly, I wish to express my sincere gratitude to my promoter Prof. Dr. ir. J. C. Schippers and supervisor Dr. M. D. Kennedy for giving me the opportunity to pursue my Ph.D. Their supervision, criticism, time, endless support and encouragement during the course of my study is greatly appreciated. The knowledge and research skills which they have instilled in me will be an invaluable asset in the future.

This work would not have been possible without financial support. To this end I would like to thank firstly, IHE, NORIT N.V. Amsterdam Water Supply and Kiwa N.V. Research and Consultancy, as partners of the INKA project which was also supported by the Dutch Ministry of Economic Affairs (Senter Grant No. MIL 93402) and in particular Dr. M. Siebel for his time and effort in co-ordinating this project which allowed me to commence my Ph.D. I also wish to thank Novem, Kiwa N.V. Research and Consultancy and Hitma B.V. for sponsoring the second part of this work and the continued financial support of IHE which allowed me to complete this work.

During the experimental phase of the project I obtained assistance from many sectors which I wish to acknowledge. I am grateful to Prof. Dr. ir. G.J. Witkamp and staff at API-TUD for their help and expertise in the barium sulphate scaling research. My thanks also to Dr. ir. J. P. van der Hoek and P. Bonné at Amsterdam Water Supply and H. Folmer, Dr. ir. J. Kruithof and H. Scheerman at Provincial Water Supply North Holland who allowed me to carry out research at their reverse osmosis pilot plants and for their patience, technical advice and support. Thanks also to G. Galjaard and Dr. B. Heijman at Kiwa for support in the MFI related research and Dr. ir. M. van der Leeden for discussions on barium sulphate scaling. The FESEM images which helped to explain my research were taken at the Electron Microscope Unit at the University of New South Wales in Australia, where Dr. M. R. Dickson generously allowed me the use of the superb facilities and Dr. H. Liang who helped me in obtaining these images. Special thanks to Fred Kruis and his staff at the Laboratory of IHE, for their support and interest in my project and to the IHE computer group and support staff for their help during my Ph.D.

Nine MSc students T. Koprowski, B. Gajda, A. Azovska, I. Bremere, M. Petros, E. Abogrean, Z. Tarawneh, D. El -Hodali, and Li Zhizhong and two stagiers R. van Emmerik and R. de Faber were involved in this study. I am sincerely grateful to their valuable contributions and collaboration. I thank you all for your patience and humour, I hope you learnt as much from me as I did from you.

My heartfelt thanks for the loyal friendship and support over the years of Niko Moerman and Martina Brosowsky, Jussi and Aija Itkonen, Mitja and Katarina Praznik, Clare McGivern and Nestor Mendez. Also, my thanks to Massimiliano Bellotti, Sonia Silva Silva, Maria Christina Miglio, Ingrida Bremere, Wendy Sturrock, Andreja Jonoski, Pere Borrull-Munt, Angela Perez, Sharyn Westlake, Tony and Nola Dicks, Martin Lopez-Velasco, Khin Ni Ni Thein, Marco Torrico Torrico, Zsuzsanna Magosanyi, Natasa and Renato Gecan, Craig Mackay, Jenny Hirst and Coniglio for your friendship, music and laughter.

Finally, I cannot thank my mother, father and brother René enough for their inspiration and for always, always being there. Also, for the phenomenal support of my extended family when I longed for south pacific shores, rhythms and the sea, with calls day and night across the world, especially from my Aunty Sheelagh, my late Uncle Alexis, Suzanne, Paddy, Thelma, Aunty Alice, Steve, Petra and Tony. Without my family's financial support, generosity of spirit, verve, kindly ear and open door, my Ph.D here would not have been tenable. Thank you.

Abstract

In the last decade, pressure driven membrane filtration processes; reverse osmosis, nano, ultra and micro-filtration have undergone steady growth. Drivers for this growth include desalination to combat water scarcity and the removal of various material from water to comply with increasingly stringent environmental legislation e.g. *Giardia* and *Cryptosporidum* removal guidelines of the Surface Water Treatment Rule (USA). Innovations in membrane manufacturing and process conditions have led to a dramatic decrease in membrane filtration costs. Consequently, membrane filtration has emerged as a cost competitive and viable alternative to conventional methods in drinking and industrial water production and in recycling and reuse. The potential of membrane filtration to solve our water quality problems is certainly only in its infancy as new applications and products emerge. However, membrane scaling, biofouling, organic fouling and particulate fouling (in this thesis scaling and particulate fouling were studied) exert severe limitations to the future growth of membrane technology. Scaling, occurring mainly in reverse osmosis (RO) and nanofiltration (NF), refers to the deposition of "hard scale" on the membrane due to the solubility of sparingly soluble salts e.g. $BaSO_4$ being exceeded. Whereas, particulate fouling is an especially persistent problem in all membrane filtration processes and refers to the deposition of suspended matter, colloids and micro-organisms on the membrane. Problems arising from scaling and particulate fouling are a reduction in product water flux or increasing operational pressures to maintain flux, which translates to increased operational costs. Membrane cleaning to remove scalants and foulants results in increased down time, energy and chemical use, and the production of waste water adding further costs attributable to fouling. Furthermore, if membrane cleaning is unsuccessful, the membranes have to be replaced to maintain production capacity.

It is widely recognised that the control of scaling and particulate fouling is instrumental in further membrane technology advancement and in decreasing costs associated with this process. However, this can only be achieved when reliable methods are available to predict and monitor the scaling and particulate fouling of feedwater and at present no such methods exist. Therefore, pilot plant operation is commonly used prior to designing full scale systems. Although this method generally provides reasonably good reproducibility, it is time consuming and expensive. The goal of this research was to develop methods to predict scaling (using barium sulphate as a model scalant) and particulate fouling in membrane filtration systems. These methods can be applied as tools to determine and monitor the efficiency of scaling and particulate fouling prevention techniques, for improvements thereof in the absence of expensive pilot plant studies and ultimately reduce costs.

Chapter 1 of the thesis gives an overview of membrane filtration in drinking and industrial water production and describes the most commonly occurring scalants and foulants and existing methods to predict and control these phenomena. Limitations of the existing methods in predicting scaling and fouling were illustrated. Whereby, at one RO pilot plant in the Netherlands treating River Rhine water, barium sulphate scaling occurred despite preventative measures i.e. antiscalant addition. While, under other operating conditions without antiscalant addition, no scaling occurred despite the high scaling tendency predicted for the concentrate. Similarly, the most widely used methods to predict particulate fouling i.e. the Silt Density Index (SDI), and the Modified Fouling Index ($MFI_{0.45}$) which simulate membrane fouling by filtering the feedwater through a $0.45 \mu m$ microfilter in dead-end flow at constant pressure, are not sensitive to the presence of smaller particles. Furthermore, the

SDI is not based on any filtration mechanism and is not proportional with particle concentration. Therefore, it can not be used as the basis of a model to predict the rate of flux decline due to particulate fouling. In contrast, the $MFI_{0.45}$ index is based on cake filtration and is proportional to particle concentration and can be used to model particulate fouling. However, it does not satisfactorily correlate with particulate fouling observed in practice as it is not sensitive to the smaller particles which may be responsible for fouling. In order to carry out the research goal of this study, scaling and particulate fouling were split into two major research branches with specific research objectives to establish (1) the solubility and kinetics of scaling and to develop an approach for scaling prediction, using barium sulphate as a model scalant and (2) an accurate predictive test to determine the particulate fouling potential of a feedwater (further development of the Modified Fouling Index making use of ultrafiltration membranes with smaller pores). This was followed by the application of these methods to determine the efficiency of scaling and particulate fouling prevention techniques.

In Chapter 2, the accuracy of the most commonly employed method for predicting barium scaling i.e. the Du Pont Manual was examined. This method predicted the barium solubility of concentrate at the RO pilot plant of Amsterdam Water Supply (AWS) was exceeded by 14 times at 80% recovery at the fixed temperature of prediction of 25°C. Yet no scaling occurred at the pilot plant for more than one year at this recovery. Possible explanations; inaccurate solubility prediction i.e. the RO concentrate were not really supersaturated and/or organic matter complexed barium were investigated. Seeded growth determination of barium solubility in the RO and synthetic concentrate (no organic matter) confirmed stable supersaturation and proved organic matter had no effect on solubility. Du Pont's method under predicted solubility by *circa* 30% at 25°C. Finally, a more accurate method was developed and verified to predict solubility (and hence quantify supersaturation) in RO concentrate in the temperature range of 5-25°C. This method uses Pitzer coefficients and an experimentally determined solubility product constant (K_{sp}) for the RO concentrate.

In Chapter 3 the cause of stable supersaturation in the AWS RO concentrate, either slow precipitation kinetics and/or inhibition of kinetics by organic matter, was investigated. Barium sulphate precipitation kinetics; crystal nucleation, measured as induction time, and growth were investigated in batch experiments in RO concentrate and in synthetic concentrate containing (i) no organic matter and (ii) commercial humic acid. Supersaturation appeared to control induction time. Induction time decreased more than 36 times with a recovery increase from 80% to 90%, corresponding to a supersaturation of 3.1 and 4.9, respectively. Organic matter in 90% RO concentrate did not prolong induction time (5.5 hour). Whereas, commercial humic acid extended induction time in 90% synthetic concentrate to more than 200 hours. This was most likely due to growth inhibition as growth rates determined by seeded growth in synthetic concentrate containing commercial humic acid were reduced by a factor of 6. In comparison, growth rates were retarded only 2.5 times by organic matter in RO concentrate. However, growth rates measured for 80 and 90% RO concentrate were still significant and not likely to limit barium sulphate scaling. Results indicate that the nucleation rate expressed as induction time is governing the occurrence of scaling.

In Chapter 4 a more realistic method was developed to predict barite scaling based on the assumption that a threshold induction time can be defined which should not be exceeded to prevent scaling. Induction times were calculated for supersaturation (determined using the Pitzer model) and temperature data from the AWS RO pilot plant from a relationship derived from measured induction times at 25°C. Safe (≥ 10 hours) and unsafe (≤ 5 hours) induction time limits, were derived from periods when scaling did and did not occur in the RO system at recoveries between 86-90%. Based on these induction times, safe and unsafe supersaturation limits were defined for 5-25 °C. Use of these limits allows more flexible operation in optimising RO recovery while avoiding scaling. The general validity of these limits should be verified in further pilot studies with feedwater of different quality and using different RO elements.

In Chapter 5, the Modified Fouling Index using ultrafiltration membranes (MFI-UF index) was developed. This index incorporates the fouling potential of smaller colloidal particles not measured by the existing $MFI_{0.45}$ or $MFI_{0.05}$ tests. In order to propose a suitable reference membrane for the MFI-UF test, polysulphone and polyacrylonitrile UF membranes of a broad pore size expressed as molecular-weight-cut-off (MWCO) 1-100 kDa were examined in tap water experiments. The measured MFI-UF (2000 - 13 300s/l^2) were significantly higher than the $MFI_{0.45}$ expected for tap water, (1 - 5s/l^2), indicating smaller particles were retained as the MFI is dependent on particle size through the Carmen-Kozeny equation for specific cake resistance. However, the MFI-UF appeared MWCO independent within the 3-100 kDa MWCO range as most likely the cake itself acts as a second membrane, determining the size of particles retained and the resultant MFI-UF. The polyacrylonitrile membrane of 13 kDa MWCO was proposed as the most suitable reference membrane for the MFI-UF test as cake filtration, the basis of the MFI test, was proven to be the dominant filtration mechanism, demonstrated by linearity in the t/V versus V plot. This results in a stable MFI-UF value over time. Furthermore, field emission scanning electron microscopy of the membrane surface showed the pores were circa 1000 times smaller than the pores of the existing $MFI_{0.45}$ test membrane.

Chapter 6 investigated various aspects of the new MFI-UF test to establish its general use for characterising the fouling potential of feedwater. Namely, proof of cake filtration and linearity of the MFI-UF index with particulate concentration of low and high fouling feedwater, reproducibility of the MFI-UF index, methods to correct the MFI-UF index for test pressure and temperature differences to the standard reference conditions of 2 bar and 20 °C, respectively and application of the MFI-UF as a monitor to detect feedwater changes over time. Cake filtration was demonstrated for high and low fouling feedwater as the MFI-UF was stable over time and proportional to particulate concentration for all feedwater tested. Reproducibility of the MFI-UF was found for 83% of the membranes tested from three different batches and in five tests using one membrane with chemical cleaning of the membrane between measurements. Correction to the reference temperature of the MFI-UF test required only correction of the feedwater viscosity. However, all the cakes formed by the filtration of the feedwaters tested were found to be pressure dependent i.e. cake compression occurred. Therefore, pressure compressibility coefficients were determined for a given feedwater and a global compressibility coefficient was calculated to correct to the standard reference pressure. At present the MFI-UF test can not be applied to quantify the fouling potential of a variable feedwater over time i.e. operate as a monitor, as the resultant MFI-UF value may be due to the combination of cake filtration with depth filtration and/or compression effects. Moreover, the delayed response in the MFI-UF index to a change in feedwater, may be due to the history effect in the calculation of the MFI-UF via the t/V vs V plot. More accurate measurement of time and volume is expected to resolve this problem and warrants further research. However, results in this chapter showed that the MFI-UF test can be used to characterise the fouling potential of a single given feedwater type and to register a change in feedwater quality.

In Chapter 7 the MFI-UF was applied to measure and predict the particulate fouling potential of reverse osmosis (RO) feedwater. MFI-UF measurements were carried out under constant pressure filtration at the IJssel Lake and River Rhine RO pilot plants of the influent feedwater and after pretreatment processes e.g. coagulation, sedimentation, conventional filtration, ultrafiltration, etc. Using the MFI-UF results, the pretreatment efficiency was evaluated and a comparison made with the $MFI_{0.45}$ which measures larger particles. The MFI-UF of the influent feedwater was circa 700 - 1900 times higher than the corresponding $MFI_{0.45}$, due to the retention of smaller particles. A pretreatment efficiency of $\geq 80\%$, was found by MFI-UF measurements at both plants. For the larger particles the $MFI_{0.45}$ gave a 90-$\approx 100\%$ reduction. Minimum predicted run times for a 15% flux decline from MFI-UF measurements, assuming cake filtration occurs in the RO systems, were shorter than that observed at the IJssel Lake plant. This was most likely due to almost negligible particle deposition in the RO systems and/or particle removal from the cake formed under cross flow. Moreover, it was shown that cake resistance increased with ionic strength in MFI-UF tap water experiments and therefore, a correction of the MFI-UF index is required for salinity effects in RO concentrate. Finally, it was suggested that the MFI-UF be carried out under constant flux (CF) filtration to more closely simulate fouling in RO systems. Preliminary experiments were promising, the MFI-UF could be determined under CF filtration within ≈ 2 hours for the low and high fouling feedwater examined and the fouling index I of the MFI-UF determined in the CF mode was linear with particulate concentration. In conclusion, the MFI-UF (measured at constant pressure or constant flux) was found to be a promising tool for measuring the particulate fouling potential of a feedwater. It can be used alone or in combination with the $MFI_{0.45}$ to compare the efficiency of various pretreatment processes for the removal of selected particle sizes and to determine the deposition of particles in target membrane systems.

In Chapter 8 the main conclusions of the research were summarised.

1

Membrane Filtration in Drinking and Industrial Water Production

Contents

Abstract

Initially, reverse osmosis (RO) was considered as an expensive luxury and applied only in arid regions for desalting sea and brackish water. Advances in membrane manufacturing and process conditions have led to a dramatic decline in costs for not only RO but also the other pressure driven membrane filtration processes; nano (NF), ultra and micro-filtration. Consequently, these processes are increasingly being applied for all water sources *e.g.* sea, ground and surface water in potable water production and in water reuse and recycling to achieve various water treatment goals e.g. removal of salts, pesticides, *Giardia* and *Cryptosporidum,* bacteria etc. Today, membrane filtration has emerged as a cost competitive and viable technology and is taking over market share from competing technologies. However, membrane scaling and fouling may exert severe limitations on future growth in membrane filtration. Scaling, occurring only in RO and NF, refers to the deposition of "hard scale" on the membrane due to the solubility of sparingly soluble salts being exceeded. Whereas, fouling refers to the deposition of various material e.g. organics, particulates and occurs in all membrane filtration processes. Particulate fouling (suspended matter, colloids and micro-organisms) is an especially persistent problem. One undesirable consequence of scaling and fouling is a decline in membrane flux (water production) or increased operating pressures to maintain flux with an ensuing increase in energy costs. Presently, no reliable parameters exist to predict and monitor the scaling and particulate fouling potential of feedwater in order to prevent these phenomena. Therefore, pilot plant operation is commonly used prior to designing full scale systems. However, although this method generally provides reasonably good reproducibility, it is time consuming and expensive. At RO pilot plants operating in the Netherlands treating River Rhine Water, barium sulphate scaling ($BaSO_4$) and particulate fouling problems occurred. At one plant despite preventative measures i.e. antiscalant addition, $BaSO_4$ scaling occurred, while under other operating conditions without antiscalant addition, no scaling occurred despite the high scaling tendency predicted for the concentrate. When scaling occurred several cleanings were required to remove it. Whereas, particulate fouling generally requires cleaning on a regular basis. The goal of this research is to contribute to the understanding of the processes involved in ($BaSO_4$) scaling and particulate fouling in membrane filtration systems, and to develop methods to predict and prevent these phenomena. These methods can be used as tools to determine the efficiency of scaling and particulate fouling prevention techniques, for improvements thereof in the absence of expensive pilot plant studies and ultimately reduce costs.

1.1   Introduction

According to the United Nations Declaration of human rights (UN 1948) every human being has the right to life sustaining resources, including water for drinking, food, industry and well being. However, 97.2% of earths water is salt water which cannot be used for drinking, agricultural and many industrial purposes [1]. Only 0.6% of earth's water resources are fresh water and due to the unequal spatial distribution of water reserves and increasing populations, the per capita availability of fresh water is decreasing [1]. Currently, 470 million people live in regions where severe water shortages exist e.g. the Middle East, Northern India, Mexico and Western US and projections indicate this may increase to 3 billion by 2025 [2]. The transport of fresh water, along with desalination and water reuse may alleviate these water shortages. In particular, the use of non conventional water resources i.e. sea and brackish water may provide a sustainable source of fresh water. However, sea water contains a high concentration of total dissolved solids 20 000 to 50 000 mg/L TDS, while the TDS content of brackish water is lower ranging from 3000 to 20 000 mg/L TDS and waters with a TDS of only 1000 mg/L are generally unpalatable to most people owing to the high sodium and chloride content. By desalination, salts are removed from sea and brackish water, lowering the TDS to potable water quality. The desalination of seawater already generates a sustainable source of fresh water in the most arid regions of the world, particularly, in the Arabian Gulf and North Africa. Without desalination, many of these regions would have remained uninhabited e.g. Kuwait and Qatar where the municipal and industrial supplies are wholly reliant on desalination. Desalination may also be the key to resolving water disputes. For instance large scale desalination is one of the proposed solutions to increase scarce water resources in the Israel-Palestine-Jordan region where water security is one of the major issues in the Israeli-Palestinian peace negotiations [3,4].

Not surprisingly then, desalination has grown markedly over the years. In 1965 the world desalting capacity was only 6000 m³/day by 2000 it had increased to 26 million m³/day [5] and is expected to double in the next 20 years representing a market increase of $US 65 billion. The major proportion of desalting capacity today remains installed in arid regions, principally Saudi Arabia (21%) followed by USA (16.8%). Seawater accounts for more than half of the desalting capacity, brackish water 25%, and the remainder comprises other sources e.g. river water. Traditionally, the desalination of seawater in the Arabian gulf region was achieved by distillation methods e.g. multi stage flash (MSF) and multi effect distillation (MED) taking advantage of the cheap energy source of oil reserves at their disposal. In these processes seawater is first evaporated and then condensed to separate it from its salt content. While MSF and MED are reliable technologies, they are energy intensive processes. Therefore, distillation plants are often coupled to power plants (cogeneration plants) to save energy. A disadvantage of this is the fluctuating demand for water and power with season and the faster growth in water demand compared to power which makes it difficult to choose and optimise cogeneration equipment [6].

An alternative desalination method is reverse osmosis (RO) which is more energetically favourable as no phase transformation is required, only electrical energy to drive the high pressure pumps to overcome the osmotic pressure of the seawater. RO is especially favourable for seawater applications with a lower salt content e.g 21 000 mg/L (North Sea) and 35 000 mg/L (Atlantic Sea) or for brackish water. Over the last thirty years RO has developed into a competitive process due to improvements in membrane technology leading to lower energy consumption, increased reliability combined with lower specific investment costs, shorter plant construction time and easy extension of plant capacity in comparison with thermal processes. Hence, in recent years the RO process has increased its market share of the world desalting capacity at the expense of thermal processes; 32.7% (1993), 35.9% in (1995), 39.5% in 1997 to 41.1% in 2000 [5].

Desalination to combat water scarcity has not been the only driver in the increased use of RO membranes. Regulatory pressure in Europe for controlling the concentration of pesticides in drinking water led to higher standards e.g. 0.1ug/l for the herbicide Atrazine. This led to RO and nanofiltration membranes being considered for the removal of pesticides and for a multitude of other treatments goals for example; removal of natural organic matter and other micropollutants, nitrate, salts and for softening. While in the USA the Surface Water Treatment Rule (1989) established treatment requirements of 3 log removal (or inactivation) protozoa *Giardia* cysts and *Cryptosporidum* oocysts, and 4 log removal of viruses [7]. These requirements might be met by increasing the dosage of chemical disinfectants e.g. chlorine. However, amendments to the safe drinking water act gave a maximum contaminant level for trihalomethanes (THMs) which are oxidation by-products formed by the chlorination of surface water containing humic acid. The increasingly stringent environmental legislation world wide thus, provided an impetus for the development of not only RO membrane research but also other pressure driven membrane processes namely nanofiltration, ultrafiltration and microfiltration in order to comply with new standards. These membrane filtration processes have differing capacities for removing targeted compounds due to their differing membrane surface characteristics. Classification of these membrane technologies based on their pore size and the size of particles and molecules retained compared to the conventional sand filter is illustrated in Figure 1.1.

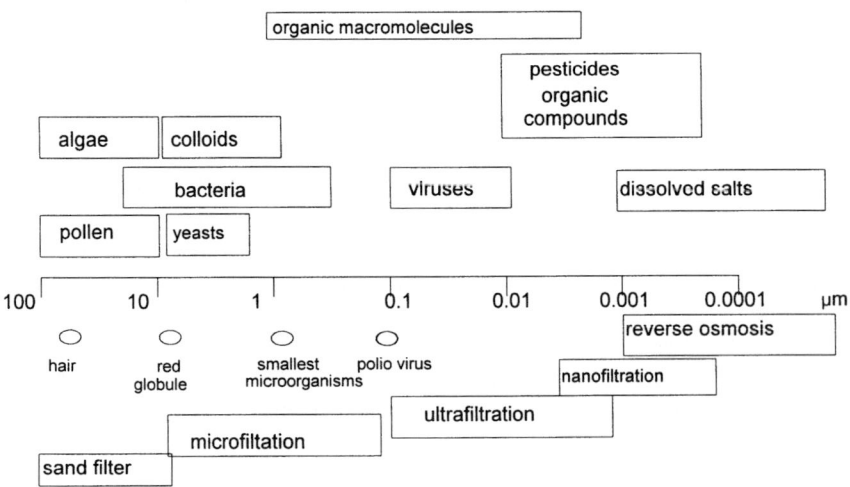

Figure 1.1:   Classification of the pressure driven membrane filtration processes and conventional sand filtration based on the size of particles and molecules removed [8].

Like reverse osmosis membranes, nanofiltration (NF) membranes do not have visible pores and remove salts via the solution-diffusion process. NF membranes can be considered as "loose RO membranes" and therefore, research carried out on RO is often also valid for NF. But contrary to reverse osmosis, nanofiltration membranes contain fixed (negatively) charged functional groups [9]. As a consequence, the selectivity of NF for monovalant and bivalent ions is significantly different. Typically, divalent ions are almost completely removed e.g. sulphate rejection is 96-98% while the rejection of monovalent ions is less e.g. for chloride 55% or lower [9]. In addition, NF may remove synthetic and natural organic compounds by size exclusion from feedwater without complete de-mineralisation of the feedwater.

On the other hand, micro and ultrafiltration are porous and separation is achieved by sieving. Microfiltration (MF) is a direct extension of conventional filtration into the sub-micron range, allowing the removal of bacteria and suspended solids. Ultrafiltration membranes have smaller pores with the pore size usually expressed as molecular-weight-cut-off (MWCO). The nominal MWCO is a performance related parameter, defined as the lower limit of a solute molecular weight e.g. dextran, for which the rejection is 95-98% [10]

Another membrane process used in drinking water treatment for desalting, which was introduced in the early 1960s some ten years before RO, is electrodialysis (ED). ED however, does not use an applied hydrostatic pressure to separate salts from salt water. Instead the driving force for this process is an applied electrical potential gradient and ions move through an ion selective membrane towards an electrode of opposite electric charge. For instance, anions such as chloride in the feedwater are attracted to the anode and pass through an anion selective membrane, but cannot pass further beyond the anode as a cation selective membrane is placed behind it. This same process occurs on the other side but for cations, for example sodium, which pass through a cation selective membrane to the cathode after which further movement is prevented by a anion selective membrane. In this way the feedwater channels are depleted of salts. However, for ED the energy consumption goes up linearly with salinity. Consequently, this method is only cost effective to desalt brackish water. Performance of the pressure driven membrane filtration processes and electrodialysis to achieve various treatment goals in drinking water treatment are summarised in Table 1.1.

In the last decade innovations in membrane technology and applications have taken place at a spectacular rate. For example ultra low pressure RO membranes which can operate at 5-12 bar were developed. While in UF and MF the process design was changed from cross flow, where only a part of the feedwater passed through the membrane, to dead end flow where all the feedwater passes the membrane. This change resulted in a spectacular reduction in energy consumption (from 5 to 0.2 kWh/m^3). Submerged UF and MF membranes were developed by Zenon called Zeeweed which operate under a slight vacuum and the feedwater passes from outside of the membrane to inside. This allows limited pretreatment of the feedwater. Most of the aforementioned examples have lead to a dramatic decrease in energy costs. Consequently, membrane filtration has rapidly become a cost competitive and viable alternative to conventional methods and has even been hailed as the most important technological breakthrough in the last decade in not only drinking and industrial water production but also in water recycling and reuse. A well known example of *indirect* water reuse is Water Factory 21 in Orange County California, where domestic waste water is treated by reverse osmosis and the potable quality reclaimed water is injected back into coastal aquifers to prevent salt water intrusion and is partly used for irrigation and drinking water as well. Similar ground water replenishment schemes are being considered in other areas in the USA e.g. Colorado [11]. More recently ultrafiltration is applied as part of a treatment scheme for *direct* potable reuse in Windhoek Namibia's capital. The direct and indirect potable reuse increases water supplies and is expected to play an increasingly important role in water scarce regions. However, psychological considerations may postpone large scale applications.

Table 1.1: Performance of reverse osmosis (RO), nano (NF) -, ultra (UF) -, and micro (MF)- filtration and electrodialysis (ED); (+) successful removal (-) no removal and (*) removal depends on MWCO or use of activated carbon.

	MF	UF	NF	RO	ED
Turbidity	+	+	+	+	
Crypto/Giardia	+	+	+	+	
Pretreatment NF/RO	+	+			
Viruses		+	+	+	
Colour/TOC		*	+	+	
Pesticides/taste/odour		*	+	+	
Hardness			+	+	
Sulphate			+	+	+
TDS			+/-	+	+
Nitrate				+	+
Fluoride				+	+
Arsenic				+	+

Other recent trends in membrane filtration are the combination of two or more membrane filtration technologies which are then referred to as integrated membrane systems. For example the application of micro/ultrafiltration prior to nanofiltration or reverse osmosis systems. MF/UF reduces the particulate fouling potential of RO/NF feedwater. This allows higher fluxes to be achieved by the RO and NF membranes and hence less membrane area is required and capital costs decrease. Two notable integrated membranes systems recently commissioned (1999) are the Mery-sur-Oise nanofiltration plant in France and the Heemskerk reverse osmosis plant in the Netherlands. The Mery-sur-Oise produces 340 000 m^3/d and employs a novel pleated microfiltration membrane as part of its pretreatment system while the reverse osmosis membrane system at Heemskerk produces 55 000 m^3/d and uses low pressure hollow fibre ultrafiltration membranes as part of its pretreatment system [12,13]. Besides pretreatment to RO/NF, MF/UF contributes substantially to the disinfection capacity of the whole system. Future projections are that ultrafiltration is expected to replace conventional pretreatment from environmental *e.g.* use of chemicals and economic considerations *e.g.* land use and investment costs [7]. Membrane technologies may also be combined with thermal desalination processes which are then referred to as hybrid systems e.g. nanofiltration with multi-stage flash [14,15].

As a result of membrane technology innovations and the increasing diversification of applications for all water types, membrane filtration has undergone rapid growth over the last decade as can be seen from Figure 1.2. Advances in membrane technology continue and its potential to solve our water quality problems is certainly only in its infancy as new applications and products emerge. Challenges for the future include further reduction in energy costs to produce water at less than US$ 0.50/m^3 and to develop membranes with even higher salt rejection capable of withstanding higher pressures.

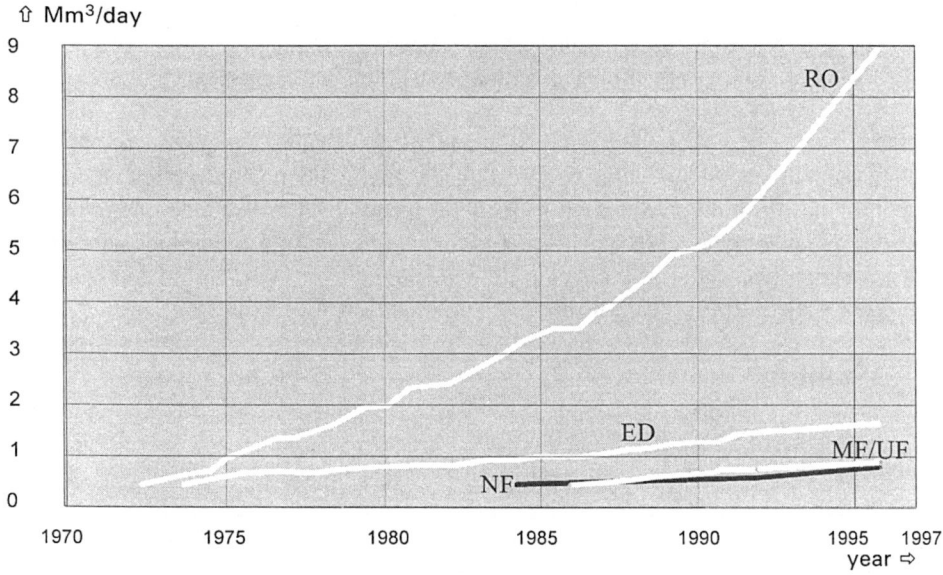

Figure 1.2: Growth in the total world capacity of reverse osmosis (RO), nano (NF) -, ultra (UF) -, and micro (MF)- filtration and electrodialysis (ED) processes [16].

However, membrane filtration is not totally the panacea in drinking and industrial water production, membrane scaling and fouling exert severe limitations on the growth of membrane technology. Scaling, occurring only in RO and NF, refers to the deposition of "hard scale" on the membrane due to the solubility of sparingly soluble salts being exceeded. Whereas, fouling refers to the deposition of various material e.g. organics, particulates and occurs in all membrane filtration processes. Particulate fouling (suspended matter, colloids and micro-organisms) is an especially persistent problem.

Undesirable consequences of scaling and fouling include; membrane failure in the worst possible case or a decline in membrane flux (water production) and in RO and NF a possible deterioration in product water quality by increased salt passage. To maintain production (flux) operating pressures are increased with an ensuing increase in energy costs. Scaling and fouling on the spacer in spiral wound elements may also cause a higher differential pressure across the spacer leading to telescoping and damage of the membrane elements. The remedy for scaling and fouling is membrane cleaning which results in increased production down time, energy and chemical use, production of waste water and a reduction of membrane lifetime which translates to further increases in costs. If membrane cleaning is unsuccessful the membranes may even have to be replaced to maintain production capacity and membrane replacement is estimated, alongside the energy costs involved in the production of water, to be the most expensive items after initial investment outlay [7].

Methods to predict membrane scaling and fouling are important tools in the control of these phenomena, both at the design stage and for monitoring during plant operation. However, presently, no reliable parameters exist to predict and monitor the scaling and particulate fouling potential of feedwater in order to prevent these phenomena. Therefore, pilot plant operation is commonly used prior to designing full scale systems. However, although this method generally provides reasonably good reproducibility, it is time consuming and expensive.

This research is concerned with methods to predict and prevent scaling and particulate fouling. As it is recognised that membrane fouling and scaling are among the most important challenges that face membrane filtration as the control of these phenomena will reduce operating and maintenance costs, membrane replacement, extend membrane lifetime and reduce energy and capital costs. Commonly encountered scalants and foulants in membrane filtration and existing methods to predict and prevent them are discussed in the following sections.

1.2   Scaling

1.2.1   Scalants

Scaling refers to the unwanted precipitation of sparingly soluble salts onto equipment surfaces during operation due to the solubility of a salt being exceeded. If the deposited solid scale layer is not removed, it tends to increase in amount over time when exposed to the same conditions. In thermal desalination plants e.g. MSF scaling represents a serious problem occurring on the heat transfer surfaces bringing about; a decrease in the volume of water evaporated, the blocking of pipes etc. Scaling is also a common problem facing the salt rejecting membrane processes i.e. reverse osmosis and nanofiltration. If the scale remains undetected and is allowed to grow into a significant scale layer it will add to the resistance of the membrane, resulting in a decline in flux at a constant applied pressure.

In the production of drinking water, brackish water and seawater have the tendency for scale formation due to the high concentrations of dissolved salts. Scalants may include; silica and divalent salts of carbonate, sulfate, fluoride, and phosphate. In seawater, the main scale forming constituents are calcium, magnesium, bicarbonate and sulfate [17]. In literature, $CaCO_3$ and $Mg(OH)_2$ are referred to as the alkaline scales because high pH favours their formation. These scales are formed by the reaction of calcium and magnesium ions with carbonate and hydroxyl ions produced by the following two reactions:

$$2HCO_3^- \rightarrow CO_2 + CO_3^{2-} + H_2O \qquad \qquad ...(1.1)$$

$$H_2O + CO_3^{2-} \rightarrow 2OH^- + CO_2 \qquad \qquad ...(1.2)$$

In addition to pH, the formation of alkaline scale is a function of the operational temperature. Calcium carbonate may form up to a temperature of 95 °C while $Mg(OH)_2$ mainly forms above temperatures of 95-100 °C due to the increase in hydroxyl formation with increasing temperature [18]. Silica has also been known to cause problems in seawater depending on the location of the plant. The chemistry of silica is more complex since it may form as a monomer, polymer, colloidal silica, or as magnesium or alumino silicates depending on feedwater chemistry and system conditions [19,20]. However, silica scaling is more often a problem in brackish ground water with higher levels of dissolved silica e.g. The Canary Islands [19]. In the desalination of brackish surface water sources, silica is less common with calcium carbonate the most commonly encountered scale and in waters with a high content of sulfate ions, calcium sulfate scaling is also common]. The presence of barium in sea and brackish water may also be a cause for concern as barium sulphate is 15 000 times less soluble than calcium sulfate. In fact it has the lowest solubility of the common scaling salts as observed from the comparison below of the solubility products (K_{sp} at 25 °C) cited in the literature [21]:

$$CaSO_4: \ 7.1\times10^{-5} \qquad CaCO_3: \ 4.96\times10^{-9} \qquad BaSO_4: \ 1.0\times10^{-10}$$

However, barium sulphate scale seldom occurred in membrane filtration until recently when applied in the treatment of River Rhine water at the RO pilot plants in the Netherlands of Amsterdam Water Supply (AWS) Company and Provincial Water Supply Company of North Holland (PWN).

1.2.2   The Mechanism of Scaling

The first requirement for precipitation or scale formation is supersaturation of the solution with respect to the salt concerned i.e. the solubility is exceeded. The maximum amount of a salt soluble in solution, at a given temperature, is given by the equilibrium *thermodynamic solubility product* K_{sp} e.g. for barium sulphate [22,23]:

$$K_{sp} = \gamma_+ [Ba^{2+}] \ \gamma_- [SO_4^{2-}] \qquad \qquad ...(1.3)$$

where $[Ba^{2+}]$ and $[SO_4^{2-}]$ are the concentration of the scalant ions in solution and γ_i is the activity coefficient of an ion and corrects for electrical interactions which occur between ions. In an ideal solution, no interaction takes place between ions and γ_i the activity coefficient is unity. The K_{sp}, and hence solubility, varies with temperature, in the case of $BaSO_4$, a temperature increase will increase its solubility [24]. However, the opposite is true for $CaSO_4$, and the alkaline scales i.e. $CaCO_3$ and $Mg(OH)_2$, which demonstrate an inverse solubility behaviour i.e. solubility decreases with increasing temperature [18].

Once supersaturation is achieved, thermodynamically, precipitation is possible. However, supersaturation is not the only aspect involved in scale formation, precipitation kinetics also plays an important role and may limit its occurrence. Precipitation kinetics comprises two steps; firstly *nucleation*; the "birth" of a new crystal of the solid phase followed by *growth* of the crystal. Either of these two processes may dictate when precipitation occurs. As supersaturation is the driving force for both processes, the higher it is the more favourable precipitation becomes [22,23]. However, a range of supersaturation may exist where precipitation may be infinitely long and solutions appear to be stable, although the solubility of a scalant is exceeded. These solutions are referred to in literature as "metastable" [22,23].

When scaling is kinetically favourable, it typically commences on an exposed surface as isolated and unconnected thin islands of crystalline solid. Other factors which may then determine the rate of scale formation are surface roughness and hydrodynamic conditions at the surface. When the crystalline solid adheres strongly to the surface and the adjacent solute layer remains supersaturated, the initial crystal islands develop by lateral and perpendicular growth into a coherent polycrystalline solid layer eventually covering the whole exposed surface. If precipitation occurs in the bulk solution the solid particles may collide with the scale layer surface, attach and cement in place by growth of the attached scale deposits [23,25]. Finally during prolonged contact of a scale with the solution, loosely bound needle like crystals or dendrites that often form the initial scale and have a rough surface, gradually transform into a smooth and tightly packed scale by recrystallisation or ageing [23].

1.2.3   Prediction of Scaling

To determine the scaling potential of concentrate, a water analysis is carried out to identify all the major anions and cations in the feedwater which may scale [26,27]. The commonly applied methods used in the membrane industry to predict $CaCO_3$, the sulphate scales and silicate are based on the Du Pont Manual and are briefly described below.

Calcium carbonate scaling is predicted using the Langlier Saturation Index (LSI) for brackish water and the Stiff and Davies Saturation Index (S&DSI) for seawater. Both indices are based on calculating the pH at which the concentrate will be saturated with respect to calcium carbonate. The LSI is defined as the pH of the concentrate minus the pH of a saturated concentrate. A negative LSI indicates no scale tendency, a LSI of zero indicates the concentrate is at equilibrium, while a positive LSI indicates a scaling tendency. The S&DSI is calculated using a similar relationship to that of the LSI while taking into account ionic strength effects in the more saline sea waters. As for the LSI, positive values indicate supersaturation and a scaling tendency, while negative values indicate no scale tendency i.e. undersaturation [26,27].

For predicting the scaling potential of the sulphate salts, the molar ionic product of each salt ($CaSO_4$, $BaSO_4$ and $SrSO_4$) is calculated in the concentrate taking into account the recovery of the system and compared with its respective solubility product. In the Du Pont Manual [27] the solubility product (K_C) used for the sulphate salts is corrected to the ionic strength of the concentrate at the desired recovery. For $BaSO_4$ and $SrSO_4$ only one K_C is reported corresponding to a temperature of $25\,^\circ C$, whereas, for $CaSO_4$ K_C ranging in temperature from $0\,^\circ C$ to $45\,^\circ C$ are reported as a function of ionic strength. When the molar ionic product of a salt exceeds its solubility product (K_C) i.e. the saturation index is greater than 1.0, this method assumes precipitation of a salt can immediately occur. To avoid precipitation, the Du Pont manual [27] recommends a recovery so that the molar ionic product is 20% below the K_C, thus ensuring a safety factor for concentration polarisation which may occur at the membrane surface which causes a higher localized concentration than in the bulk solution.

The Du Pont Manual, as for the sulphate scales, predicts SiO_2 scaling on the basis of whether the concentration of SiO_2 in the concentrate will exceed the solubility value of SiO_2 at a given temperature and pH given in literature. However, SiO_2 scaling has been observed in reverse osmosis permeators despite a silica scaling index of only 0.6-0.8 i.e. the concentrate is undersaturated with SiO_2 [28]. Butt et al attributed this to a catalytic effect by aluminium and iron, the latter of which was present at high concentrations, in the concentrate. According to the Du Pont Manual and Darton [19] the solubility of SiO_2 may decrease drastically in the presence of trivalent cations such as aluminium and iron due to the formation of sparingly soluble mixed silicates e.g. alumino silicates. Due to the aforementioned reasons and as silica has a number of different polymorphs each with different structures and solubilities, the prediction of silica scale and which form it precipitates as, remains difficult [19].

1.2.4   Scaling Control

If scaling is predicted to occur in a membrane system, various approaches may be taken to control scaling. As scalants may be difficult to remove the prevention of scale formation is the most desirable approach. This can be achieved by two main techniques, broadly categorised into physical or chemical methods which are further described below. If these methods are unsuccessful and scaling does occur, its immediate removal is desirable by methods described further in section 1.2.4.3.

1.2.4.1   Physical Scale Prevention Methods

The most simple method to prevent scaling is to lower the recovery of the system beneath the solubility limit of the salt in the concentrate. However, this is seldom applied to control scaling as it is generally uneconomical with less product water produced while the amount of concentrate requiring disposal increases. Secondly, the hydrodynamic conditions at the membrane surface can be optimised to promote turbulence at the membrane surface to prevent attachment of crystals and minimise concentration polarisation at the membrane surface and hence the chance that the solubility will be exceeded at the membrane. This typically involves increasing the cross flow velocity across the membrane surface and the use of a membrane spacer to promote turbulence. Although, in some cases the feedwater may already be saturated with the sparingly soluble salts and therefore, these methods will have a limited effect.

Additional physical methods include the application of a magnetic and/or electrical field designed to disrupt scale formation. However, contradictory results have been found where in some cases scale is prevented and in other cases enhanced [23,29,30]. Furthermore, most studies have been carried out for the prevention of calcium carbonate scale in the treatment of boiler feed water and not membrane filtration. The mechanism by how scale is prevented remains unclear. One mechanism suggests that nucleation or the formation of crystals is prevented by increasing the dielectric or magnetic permeability of the crystallizing phase above that of the supersaturated solution [23]. As the effectiveness of this technique has shown inconsistent results and the exact mechanism of its action remains unclear, this method has not often been applied for the prevention of scaling in drinking and industrial water production.

Alternatively, the scalant ions may be removed from the feedwater by passing feedwater through an ion exchange column. This is used for calcium carbonate where the total hardness is less than 300 mg/l as $CaCO_3$. Ion exchange may also be applied to remove sulfate anions from the feedwater. A more promising technique for barium sulphate scaling prevention, which is under development, is to desupersaturate the membrane concentrate by passing it through a column reactor containing barium sulphate seeds placed before the last stage of a membrane array [31,32]. As the concentrate passes through the reactor, barium and sulphate are removed from the concentrate as they adsorb onto the seed crystal surface and become incorporated into the growing seed crystals [31,32]. In so doing the concentration of barium and sulphate is reduced and may decrease down to the solubility level. Similarly, silica may be removed from the feedwater by desupersaturation and is also currently under investigation [20].

A further interesting development to prevent scaling in seawater RO (SWRO) and MSF desalination is the hybrid combination of nanofiltration with SWRO and/or MSF [14,15]. Nanofiltration applied as a pretreatment step will remove scaling ions, the NF permeate is then fed to the SWRO and/or the MSF system. This allows higher temperatures and recoveries to be applied in the MSF and SWRO processes, respectively. Consequently, a higher capacity can be achieved for an existing plant.

1.2.4.2   Chemical Scale Prevention Methods

Chemical methods to prevent scaling involve the addition of acid or acid-forming materials and antiscalant to the feedwater. Acid addition is the most common method to prevent alkaline scale formation in membrane filtration systems. Through acid addition, normally sulphuric (H_2SO_4) or hydrochloric acid (HCl), the bicarbonate ion is converted to CO_2 as follows:

$$H_2SO_4 + 2HCO_3^- \rightarrow 2H_2O + CO_2 + SO_4^{2-} \qquad \text{...(1.4)}$$

$$HCl + HCO_3^- \rightarrow H_2O + CO_2 + Cl^- \qquad \text{...(1.5)}$$

Enough acid is added to adjust the pH so that a zero or negative LSI or S&DSI is obtained and the threat of scaling is removed. Of the two acids, sulphuric acid is more commonly applied as it is normally the least expensive acid available. However, the use of H_2SO_4 leads to corrosion problems and its addition provides a source of SO_4^{2-} ions, which will add to the sulfate ion "load" in the system, increasing the potential for sulfate scale precipitation.

Although, acid addition is effective in alkaline scale prevention, it is not effective for hard sulphate scales such as barium or calcium sulfate. Therefore, to prevent the precipitation of these scales antiscalants (scale control additives) are commonly applied. Antiscalants inhibit scaling by one or more of the following mechanisms; threshold effect which keeps scalant ions in solution, crystal distortion by adsorbing onto the crystal surface at active sites preventing further growth, and finally by dispersancy where a surface charge is added onto the crystals so that they repel one another, reducing their ability to adhere to each other and equipment surfaces, hence, they remain in solution. [18,26,33].

The first antiscalants, developed in the 1930s, were polyphosphates $[O - P(O)_2 - O]$ such as sodium hexametaphosphate SHMP. SHMP was found to retard the precipitation of calcium sulfate and also calcium carbonate [34]. Thus, less acid can be dosed and a LSI of 1.0 can be tolerated [19]. However, the problems associated with SHMP are well documented. SHMP was found to be unstable, the O-P chain hydrolyses after a short period of time to form orthophosphate, which is inactive as an antiscalant. Moreover, the high concentration of phosphate in concentrate may cause problems in concentrate disposal as it enhances eutrophication.

Of particular concern in drinking water applications, the antiscalant needs to comply with health and safety standards and meet regional disposal requirements for concentrate disposal. To meet this latter requirement, antiscalants need to be biodegradable. However, this increases their biofouling potential and may generate additional problems with biodegradation byproducts (35-36). Furthermore, antiscalants may cause organic fouling through adsorption onto the membrane. Both biofouling and organic fouling of membranes will result in increased operational pressures to maintain the desired flux as discussed previously.

Over the years many materials and combinations have been suggested and trialed as antiscalants to meet all the aforementioned requirements. For instance polyphosphonates $[O_3 P- C]$ have a more stable C-P bond and are less likely to hydrolyse even at high temperatures and allow operation at higher LSI values e.g. Perma Treat 191 allows an LSI of 2.6 [17,26,28]. Moreover, they also sequestrate iron salts and are effective in the dispersal of suspended and colloidal matter [19,37]. Although, recently developed phosophonates are reported to also inhibit silica scale [38] this has not been widely confirmed. Therefore, physically removing silica by desupersaturation could be an attractive method to prevent this type of scaling. The effectiveness of antiscalants is greatly affected by the nature of the functional group, concentration, and the molecular weight of antiscalant [39]. Diphosphonates and polycarboxylates $[CH_2-CH-COOH]_n$ containing at least several anionic functional groups per molecule have been found to be the most effective inhibitors presently known to prevent barium sulphate scaling [33,34,40].

The supersaturation and temperature occurring during operation will also affect the efficiency of antiscalants. Hence, in addition to the choice of antiscalant, the dose needs to be optimised. Overdosing of certain antiscalants may enhance sludge formation while underdosing may lead to scale formation [17]. Furthermore, optimising the dose will save on antiscalant costs.

Antiscalant addition offers many advantages over acid addition. Antiscalants are used in sub stoichiometric amounts (ppm range) and therefore, do not require large quantities and storage facilities. In contrast, acids work in stoichiometric amounts and as they are corrosive and hazardous they require special storage and handling procedures on a large scale. Consequently, antiscalants such as polyphosphonates which are also effective at higher temperatures have become the preferred option in MSF and MED applications instead of acid to prevent calcium carbonate scaling. In the future this trend can be expected to be followed by RO and NF applications.

1.2.4.3   Scale Removal

In the case that scale control methods fail and scaling occurs in a system, the scaled surface needs to be cleaned. This results in the shutdown of a plant and increased operational costs accrued from energy and chemicals involved in cleaning. Effective cleaning agents may be the same as those used to prevent it from occurring [41]. For example calcium carbonate scale can be readily removed by flushing with hydrochloric acid (pH = 2.5) or with a 2 % solution of citric acid at pH = 4.0. Alternatively, ethylenediaminetetraacetic acid (EDTA) can be applied for cleaning as it has been shown to effectively dissolve calcium sulfate and calcium carbonate scales [27,42].

In membrane filtration scaling typically starts in the final element of the last stage where supersaturation is highest. However, it may not be detected immediately depending on the sensitivity of flow measurement equipment. Thus, a significant amount of scale may be formed before a decrease in flux or increase in operating pressure is observed. During this time the scale may have aged into a harder and more compact form (refer section 1.2.2.) and may be virtually impossible to remove. For instance silica scale is notoriously difficult to remove especially as it may become more viscous and dehydrate on the membrane surface to become hard like cement [43]. Similarly barium sulphate scale can be insoluble if it is allowed to age. On the other hand, if the scale is formed in the presence of antiscalants it may be softer due to crystal distortion effects and weakly attached to the membrane and then membrane cleaning may be successful.

Partially successful cleaning decreases the membrane life time or may even mean the total loss of a membrane element. In one study the life span of RO permeators was shortened by scaling problems from the guaranteed life span of 5 years by more than half to 2 -2.5 years [28]. Since membrane replacement is responsible for a substantial part of the operation cost this is undesirable. Therefore, good scale prevention should be ensured in RO and NF applications, to lower the occurrence of membrane replacement.

1.2.5   Barium Sulphate Scaling at AWS

Barium sulphate scaling was found to be a problem at the RO pilot plant of Amsterdam Water Supply (AWS) company from November 1993 to April of 1994. Initially, the pilot plant was operated at 90% recovery and sulfuric acid was added to prevent alkaline scaling. According to the method in the Du Pont Manual, the feedwater, pretreated River Rhine water, entering the RO units was already 1.5 times saturated with $BaSO_4$ (before acid addition) and scaling was expected to occur. Therefore, an antiscalant was added, nevertheless scaling occurred in the last stage, where the concentration of $BaSO_4$ was up to 70 times the saturation level. A SEM taken of the membrane surface during this time shows heavy barium sulphate scaling (refer Figure 1.3). Changing the antiscalant prevented scaling, however, biofouling/organic fouling occurred due to the biodegradability/adsorption of the antiscalant. To avoid these problems the operating conditions of the pilot plant were changed. The recovery was lowered to 80% and hydrochloric acid replaced sulfuric acid to reduce the load of sulphate ions. No antiscalant was used and no scaling was found despite the high supersaturation (14 times the saturation value) during more than 19 months of operation under these conditions.

Similar barium sulphate scaling problems were found in 1997 at another RO pilot plant operating in the Netherlands using pretreated River Rhine water, that of Provincial Water Supply Company of North Holland (PWN). The RO pilot plant was operated at 80% without antiscalant addition, in this case barium sulphate scaling was observed. However, the sulphate concentrations in the feedwater is higher than at AWS and hence the supersaturation (17 times the saturation value at 80% recovery). Moreover, biofouling also occurred which may have promoted scale formation in the system.

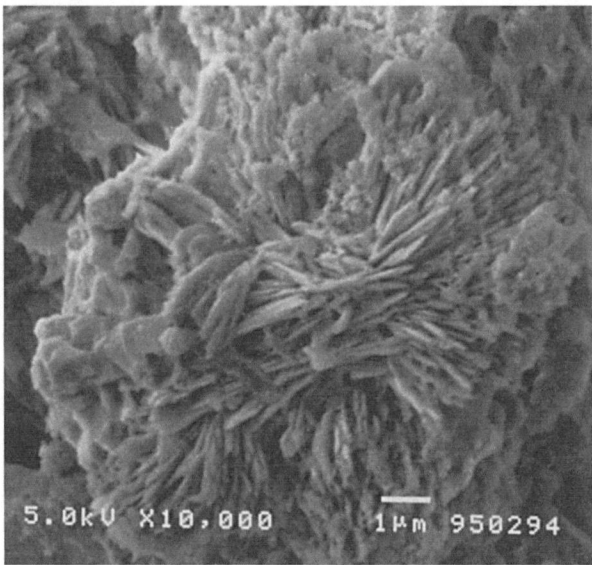

Figure 1.3:   SEM of a reverse osmosis membrane taken from the final stage of the RO pilot plant at AWS showing the presence of barium sulphate scale which occurred during operation at 90% recovery with sulphuric acid and antiscalant addition.

Surprisingly, barium sulphate scaling was not found to be a common problem before in membrane filtration. This may be attributed to the fact that until recently RO was not normally applied in treating river water. However, as surface water sources are increasingly being treated by RO the incidence of this type of scaling may increase as barium may be present in many surface waters from both natural and anthropogenic sources. The mineral barite occurs naturally in sedimentary rocks and is mined in the United States, England, and Germany as the main source of barium sulphate for the chemical industry. Amongst other commercial uses barium sulphate is commonly applied in the manufacture of rubber products, glass, paints as a pigment extender and as a filler in the paper industry. A distinguishing feature of barium sulphate is its high specific gravity (4.5) which has led to its use in the petroleum industry as a densifying agent in drilling muds added to obviate the danger of high gas pressures during drilling. Hence, barium sulphate scaling has long been known as a notorious problem in the extraction of oil and gas from the sea and is the principal scale found in North Sea offshore oilfields and a common scale found in the Arabian Gulf region [44]. Consequently, most research carried out to date on barium sulphate scaling is within this context, while in the context of membrane filtration not much research has been done. Moreover, the research of barium sulphate scaling in the petroleum industry is not transferable to RO and NF applications as the temperature and pressure of "down-hole" conditions are much higher.

The barium sulphate scaling problems experienced at Amsterdam Water Supply company illustrate the drawbacks associated with the existing method in predicting the scaling tendency of a feedwater. Inaccurate scale prediction using this approach may lead to an overestimation of the scalants solubility, possibly resulting in scaled membranes. Alternatively, underestimation of the solubility of the scalant may lead to an overdose of the antiscalant. However, in practice less scale than predicted occurs and in some cases no scale at all was found when a scaling tendency was indicated. Reasons for the contradictions between predicted and actually encountered scale are discussed below taking into account the mechanism for scale formation described in section 1.2.2.

Using the method from the Du Pont Manual the solubility of barium in the AWS feedwater is exceeded *i.e* the feedwater is supersaturated and scaling is already predicted. However, the feedwater contains both organic and inorganic ions. The barium solubility may be increased by the presence of one or both of these components in the feedwater matrix. As the feedwater passes through the RO system the inorganic and organics are increasingly concentrated, and consequently the solubility (and equivalent supersaturation) will change. To account for the increasing salinity the Du Pont method uses a K_C corrected for ionic strength. A more accurate estimation of solubility involves the use of the *thermodynamic* solubility product (K_{sp}). The K_{sp} is calculated using activity coefficients (γ_i) of ions which correct for ionic interactions occurring in solution at ionic strengths greater than 0.001M. There are a few ways of calculating activity coefficients. The Debye-Hückel model (1923) only takes into account long range interactions occurring between ions and is valid only to ionic strengths of 0.01M. The Bromley correlation which is an extension of the Debye-Hückel is accurate up to higher ionic strengths of 6M but also only takes into account long range interactions. Whereas, Pitzer (1974) and co-workers developed a model to calculate activity coefficients which includes long and short range interactions and may therefore give more accurate solubility prediction [45]. Alternatively, the presence of organic matter in surface water e.g. humic acid may form complexes with metal ions such as barium and calcium, thus removing the scalant ions from solution and increasing the solubility of a salt [46]. Hence, due to the inaccuracy of solubility prediction and/or organic complexation of barium the AWS feedwater and concentrate may not really be supersaturated with barium sulphate. Furthermore, RO feedwater temperature will influence the solubility of a salt and the method in the Du Pont manual is limited to 25 °C.

If the feedwater and concentrate really are supersaturated, precipitation kinetics, which are neglected by the current approach in predicting barium sulphate scaling, may limit its occurrence.

Nucleation and/or growth of barium sulphate in the concentrate may be very slow. This may be influenced by the composition of the concentrate e.g. barium and sulphate ion stoichiometry, higher ionic strength and the presence of organics. In most surface waters the sulphate ion is present far in excess of the barium ion, for example in the River Rhine up to 1000 times greater, and this may inhibit either nucleation and growth. Furthermore, organic matter of both natural and anthropogenic origin in surface water may inhibit precipitation acting as antiscalants resulting in stable supersaturation. For instance lignin, tannic and humic acid derivatives and other natural polymers from plants, have been used in scale control [41]. Lignosulfonates are classified as anionic polyelectrolytes and are used as dispersants and crystal modifiers in the molecular weight range of 1000 to 20000 [41].

However, as yet the influence of these aspects on barium sulphate scaling or other scalants have not been researched in the context of membrane filtration.

1.3   Fouling

Fouling occurs in all membrane filtration processes i.e. reverse osmosis, nano, ultra and microfiltration. Whereas, in the competing desalination processes, MSF and MED processes are very forgiving of influent feedwater quality and are less prone to fouling.

1.3.1   Foulants

Membrane fouling refers to the deposition of material onto the membrane surface causing a decline in flux over time when all operating parameters, such as pressure, flow rate, temperature and feed concentration are kept constant. Fouling may also be accompanied by an increase or decrease in salt passage. Various types of fouling can be distinguished depending on the material deposited; biological (biofouling), organic, (hyd)oxide and particulate fouling and are briefly described below.

There is a significant difference between organic and biological fouling. Biological fouling (or biofouling) is a result of microbial (bacterial/algal/fungal) attachment to the membrane and subsequent growth with the release of biopolymers as a result of microbial activity. Biological fouling may arise from sulphate reducing and anaerobic bacteria present in the raw water source, algae growth stimulated by light, and microorganisms embedded in the membranes or modules [47]. The possible degradation of the membrane material (polymer) providing a source of carbon and energy, and the presence of assimilable organic compounds (AOC) in the feedwater will promote biofouling. Organic fouling on the other hand is often taken to imply the chemical or physical adsorption of organic compounds onto the membrane which may be followed by the build-up of a cake or gel layer at the membrane surface. While there is a recognisable connection between biological/organic fouling, they should be both monitored and controlled separately.

The low concentrations of metals in water are often ignored as potential foulants, although sea and river water analyses have shown the existence of several trace metal species, such as iron, aluminium, manganese, copper, zinc, chromium and lead. However, the lowering of the pH to 7 or less during pretreatment may convert some of these metals into insoluble oxides and/or hydroxides. [48]. These compounds may then be deposited onto the membrane surface, especially by concentration polarization effects causing fouling (and are also known to deposit on heat exchange surfaces) [38]. The (hyd)oxides of metals *e.g.* $FeOH_3$, MnO_2 may deposit alone or complexed with organic or colloidal material and may grow in size by polymerization and cross linking with organic and inorganic polymers to form amorphous gelationous deposits [49]. The most common metal fouling in both sea and surface water plants is caused by iron. Iron contamination in many cases is from the pretreatment plant itself either due to corroded pipe works particularly with low pH acid dosing systems or with overdosing of ferric chloride when used as a coagulant.

Particulate fouling, the focus of this research, may be defined as the build up of particulate material *e.g.* suspended solids, colloids and microorganisms on the membrane surface. Particulate fouling, also referred to as colloidal fouling, is one of the most persistent types of fouling. Surface water from the sea (surface intake), rivers or lakes have a higher concentration of suspended solids and colloids and are especially prone to particulate fouling. In contrast, brackish waters drawn from deep wells normally have a lower concentration of particulates and therefore, particulate fouling is less of a problem for these sources.

The particulate inorganic matter in aerated river and lake waters is predominantly comprised of clay minerals and (hyd)oxides of aluminium, iron, manganese and silica [50,51]. While fulvics, humics, polysaccharides and proteins make up the organic fraction of particulate material. In the River Rhine water, which is the raw water source of both the PWN and AWS RO pilot plants, the greatest proportion of suspended matter was found to be clay minerals (mainly illite) 45%, silica 27% and organic matter 18% with carbonates making up the rest 9% [52]. On storage, the organic matter fraction was found to increase up to 61% in the IJssel lake where PWN abstracts the water to store in a reservoir before being treated. Additional sources of colloidal matter in systems may arise from corrosion products from carbon steel pumps, piping and filters prior to the membrane filtration system.

The persistence of particulate fouling, especially colloidal fouling, in membrane filtration systems can be attributed to their small size and charge. Colloidal particles are very small, ranging in size from 0.001 to 1μm. In addition they are electrically charged which is due to the presence of charged functional groups on the surface of the colloid itself or through the adsorption of ions from the surrounding water. In surface water conditions normally colloids are negatively charged and are surrounded by a diffuse double layer in water, a polar medium. Individual particles are prevented from coming into close contact with each other by the repulsive action of their negative surface charges and the double layer. Hence, colloidal particles are stable in solution for long periods of time e.g. the settling velocity of a 1nm spherical particle is 3m/million yr [53].

1.3.2   The Mechanism of Fouling

The permeate flux J [m³/m²s] through a membrane can be described by:

$$J = \frac{\Delta P}{\eta\, R_m}$$ ...(1.6)

where ΔP [N/m²] is the applied transmembrane pressure , η [Ns/m²] the water viscosity and R_m [1/m] is the resistance of a clean membrane which is a function of properties such as membrane thickness, pore size and porosity. The first step in fouling is transport of the foulants to the membrane surface. In dead end mode all feedwater passes through the membrane and particles are carried to the surface of the membrane. In cross flow the feedwater flows tangentially across the membrane surface with only a part of the feedwater passing through the membrane. Deposition on the membrane surface in this mode depends on the forces acting on the particle and its size. At the membrane surface foulants may become attached to the membrane by processes such as adsorption, precipitation or convectively-driven plugging [53]. Four major mechanisms are commonly used to describe how foulants deposit and are explained in terms of conventional filtration theory, depicted in Figure 1.4, where particles may (1) completely seal a pore referred to as complete blocking, (2) enter into a pore restricting the pore volume called standard blocking, (3) seal a pore and form a surface deposit called intermediate blocking or (4) simply form a surface cake deposit i.e. cake filtration [54].

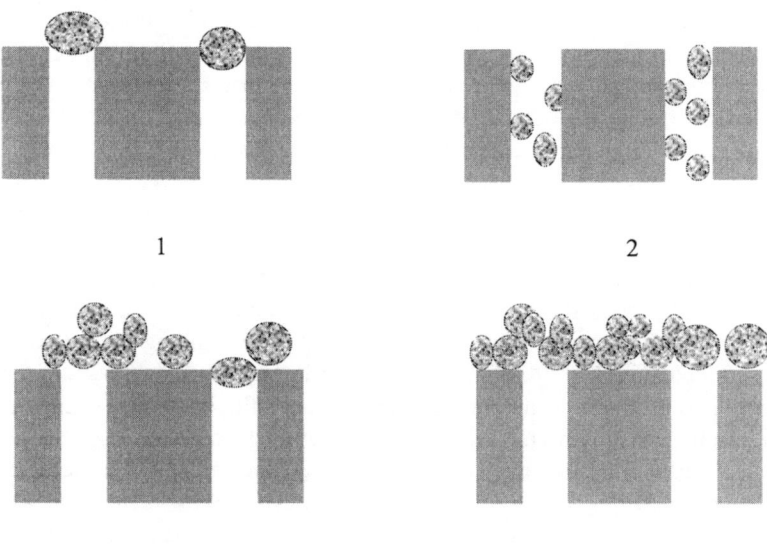

1   2

3   4

Figure 1.4:   Schematic representation of membrane fouling by conventional blocking (1) complete, (2) standard and (3) intermediate and cake (4) filtration mechanisms [adapted from 53]

The blocking of a pore (R_b) or the formation of a cake (R_c) adds further resistance to the membrane resulting in a decrease in flux via Equation 1.7.

$$J = \frac{\Delta P}{\eta (R_m + R_b + R_c)} \qquad \qquad ...(1.7)$$

The above models are valid for all types of fouling and also scaling. However, scaling first involves a phase change of the scalant ions into a crystalline solid either in the concentrate or on the membrane surface. As filtration proceeds cake filtration is expected to be the dominant fouling mechanism and also in the case of biofouling where biomass is surrounded by polymers.

1.3.3   Fouling Control

Membranes represent a significant part of plant costs, thus, maintenance of good membrane performance must be a prime objective of any membrane filtration system. Tactics to control membrane fouling take one of following approaches: preventing fouling from occurring by pretreating the feedwater to remove foulants and promoting conditions at the membrane surface to prevent their deposition, removing fouling by periodic membrane cleaning and manipulation of the filtration mechanism to improve cleaning efficiency.

1.3.3.1   Fouling Prevention

The complexity of the pretreatment system is dictated by the quality and consistency of the source water, membrane material and configuration and recovery. Minimum pretreatment is generally required for brackish well water (depending on the iron content) since the water has already been filtered through the earth. In this case pretreatment may consist of only a 10μm cartridge filter to remove large particulates, which may clog the membrane.

Surface water is normally more variable due to seasonal rainfall and temperature changes e.g. algal blooms and the higher concentration of suspended solids and organic compounds. Therefore, these sources require a higher level of pretreatment especially in RO or NF applications which may consist of conventional and/or advanced pretreatment processes. Conventional pretreatment consists of the commonly employed drinking water treatment steps of coagulation, sedimentation and filtration using sand and/or multi media filters. In these filters particles found in the raw source water are agglomerated and flocculated by chemicals such as ferric chloride, alum and polymers. Extended pretreatment processes e.g. ozonation and biological activated carbon filtration may also be applied and are part of the pretreatment system at the RO pilot plant of AWS. The combination of these two pretreatment steps will remove organic compounds e.g. humic acids and may then lower the incidence of organic fouling. Also, as mentioned previously, UF and MF are increasingly being applied prior to RO as part of an integrated membrane system. Such a system was successfully trialed at the PWN RO pilot plant and now UF is used in their full scale plant as part of the treatment process prior to RO.

As in scaling (refer section 1.2.4.2.) antiscalants/antifoulants may be added to the feedwater which disperse colloidal particles and may reduce particulate fouling and reduce iron (hyd) oxide fouling as they sequester iron as well. Physical methods in the membrane system to prevent attachment of foulants to the membrane surface e.g turbulence promoters, smooth surfaces and wider spacers (recently developed by Dow) are also similar to those mentioned for preventing scaling described in section 1.2.4.1.

1.3.3.2   Fouling Removal

Complete removal of foulants is in general impossible on practical and economic grounds. Therefore, membrane cleaning is an essential component in reducing membrane fouling. Cleaning methods and agents are specific not only for the type of foulant, but also for different membrane materials and the geometric configurations in which they are used.

For dead end MF and UF systems frequent hydraulic cleaning (backwashing) is applied e.g. every 15 minutes for 15 seconds. Disadvantages of this are shorter run times and it consumes valuable product water. Moreover, after some time chemical cleaning is required e.g. every 8 hours as the efficiency of backwashing decreases over time. To avoid excessive use of chemicals, product water and down time, cross flushing intermittently with feedwater has been under investigation [56]. Results showed that the success of cross flushing depends on how the foulant is deposited. Cross flushing is expected to be more successful for removing foulant deposited as a cake. Whereas, foulants blocking pores will be more easily removed by backwashing. Therefore, further research focuses at optimising cross flushing and backwashing to give effective and efficient system performance in terms of cleaning and costs [56,57].

Recently, novel applications are under investigation to manipulate the filtration mechanism occurring in UF and MF systems. One promising system, enhanced pre-coat engineering (EPCE) developed by Galjaard [58], involves the prior deposition of ferric hydroxide flocs to form an easily removable, permeable and incompressible cake layer on the membrane. This protective cake prevents pore blocking and traps colloids, algae and other suspended matter. When the applied transmembrane pressure increases above a certain value the membrane is backwashed and the cake is easily detached. The fouling rate was decreased dramatically and stable operation at a higher flux was demonstrated [58].

Backwashing is not possible for spiral wound RO and NF configurations. Therefore, the system needs to be shut down for cleaning (normally chemical) which is commonly recommended when a 15-20% decrease in the normalised flux or increase in pressure drop of an installation is observed. Disadvantages of cleaning are it is costly in down time, energy, and in the chemical cleaning agents applied and for the disposal of the polluted wastewater. Moreover, depending on the nature and extent of fouling, only partial flux restoration may be achieved or powerful cleansing agents may be required which may damage the membrane. Typically, cleaning yields diminishing returns the longer the membrane is used. This loss in flux over time is usually considered in the design of the plant by increasing the surface area accordingly.

In the worst case, cleaning is unsuccessful and the fouling is irreversible, then membranes need to be replaced which makes membrane filtration a very costly operation. Therefore, adequate pretreatment should be ensured especially in the case of RO and NF systems.

1.3.4   Fouling Prediction

The modelling of fouling is complex due to the variety of foulants and their synergistic interaction with each other and the membrane. At present no reliable methods exist by which the rate of fouling can be accurately predicted. However, useful indices for fouling prediction of a feedwater are necessary for process design, for the determination of the required feedwater pretreatment and for predicting flux decline. This latter aspect is important in estimating operational costs associated with pretreatment and also membrane replacement or cleaning.

For the prediction of the biofouling potential of feedwater, no model exists which relates flux decline and the occurrence of biofouling of a feedwater. Current methods used to predict the biofouling potential such as standard plate counts, will in reality only give an indication of the presence or absence of bacteria and not the potential for attachment to the membrane or the rate of growth. The measurement of parameters such as assimilable organic carbon (AOC) or biofilm formation potential, would give a better prediction of the potential of biofouling. The latter of which is currently under investigation at PWN. For the prediction of organic fouling no parameter is available at present. While the prediction of metal (hyd) oxide fouling is based on measuring the concentration of metal ions in feedwater.

Water quality parameters such as turbidity, suspended matter and particle counting can be used to indicate the concentration of particles in a feedwater. In particular, particle counting has proven particularly successful in monitoring particle removal and the integrity of ultrafiltration membranes [59-62]. However, the aforementioned parameters cannot be used to estimate the particulate fouling potential of a feedwater arising from the deposition of particles on the membrane or spacer. Existing methods designed to measure the particulate fouling potential of feedwater are; the Silt Density Index (SDI), and the Modified Fouling Index (MFI). These tests are designed to simulate the membrane fouling process.

The Silt Density Index (SDI) is the most commonly employed method to predict particulate fouling. The SDI was introduced by the Du Pont Company and employs a 0.45µm microfiltration membrane filter to empirically predict fouling of their RO hollow fine fibre permeators by a particular feed water. The SDI is a valuable empirical test for these permeators in that it gives information on the plugging of the non-woven material (pore blocking plus cake filtration mechanism) together with the bundles of the RO fibres (depth filtration mechanism). For spiral wound modules, the SDI is also applied, however, for these modules the filtration mechanisms are not exactly the same as that which operate in a hollow fine fibre permeators, therefore the test can not be directly translated for fouling predication of spiral wound modules.

The SDI test consists of passing the feedwater through the 0.45µm microfiltration membrane under a constant applied pressure and determining the filter plugging rate. The filter-plugging rate is determined by measuring the time to collect the initial sample filtered through the membrane, continuing the filtration for up to 15 minutes, and then measuring the time to collect the second sample [27]. The sample volume collected is usually 500 ml. The SDI is calculated from the following equation:

$$SDI = \frac{100\left[1 - \dfrac{t_i}{t_f}\right]}{t_t} \qquad \qquad ...(1.8)$$

where t_i is the time to collect the first 500 ml of water which passes through the filter, and t_f the time taken to collect an additional 500 ml after a filtration time t_t of 5, 10 or 15 minutes (usually 15 minutes). Most membrane manufacturers use the SDI results as the basis for their membrane guarantee. Rule of thumb guidelines for the SDI, are an SDI < 1 is preferable for hollow fine fibre with a maximum SDI of 3 and for spiral wound RO membranes a maximum SDI of 4-5 is preferable.

However, despite the wide acceptance of the SDI the test is a poor simulation of actual RO conditions. The SDI is a static measurement of resistance which is determined by samples taken at the beginning and at the end of the test. The SDI makes an intrinsic assumption that the flows at these two instants are linearly related. For fairly clean waters, the initial and final flow are similar and the SDI's convention of assuming a linear approximation does not give misleading results. However, highly fouling waters generally do not result in a linear flow reduction since the resistance of the cake being deposited upon the filter pad is a function of volume instead of time [63]. Moreover, the SDI does not measure the rate of change of resistance during the test and makes no distinction between filtration mechanisms *i.e.* pore blocking, cake formation and cake compression. Furthermore, there is no linear relationship between the SDI index and the concentration of colloidal and suspended matter with the index limited mathematically to a maximum value of 6.66, 10 and 20 for a filtration time of 5, 10 or 15 minutes, respectively [64]. As the SDI has no theoretical basis it is not possible to model flux decline in RO systems from the SDI.

The Modified Fouling Index ($MFI_{0.45}$) was derived by Schippers et al. [64-67] from the Silt Density Index and is determined using the same pore size membrane as for the SDI and under a constant applied pressure. Unlike the SDI the permeate volume is recorded every 30 seconds over the filtration period. Moreover, the $MFI_{0.45}$ is based on the cake filtration mechanism (refer section 1.3.2) and a linear relationship between the $MFI_{0.45}$ index and the concentration of colloids and suspended matter in feedwater has been demonstrated [64]. Therefore, the MFI can be used as a basis for modelling flux decline in RO (and NF) systems. Assuming that particulate fouling on the surface of reverse osmosis (or nanofiltration) membranes can also be described by the cake filtration mechanism, a MFI model was developed to predict flux decline or pressure increase to maintain constant capacity in RO systems [65-66].

The $MFI_{0.45}$ is mainly used in the drinking water industry for the prediction of membrane fouling but has also been found to be a good parameter to estimate the rate of clogging of artificial recharge wells [66]. However, in membrane filtration applications no sound correlation between the $MFI_{0.45}$ and the fouling of the actual RO membrane was found. Experimental research carried out at the RO pilot plant of PWN by Schippers when the feedwater was pretreated by in line coagulation resulted in $MFI_{0.45}$ values below 1.5 s/l^2. According to the MFI model 15 % flux decline should occur in the spiral wound RO membrane system after 274 years for $MFI_{0.45}$ values in this range. However, in practice 15 % flux decline was reached after about 2 months. Fouling due to colloidal particles less than 0.45μm was expected to be responsible for this deviation. Measurements of the MFI value measured for the RO feedwater using membrane filters of varying pore size in the range of 0.015 - 0.8μm demonstrated that the MFI value sharply increased with decreasing pore size, as given in Table 1.2. This confirmed that the water contained many small particles that most likely passed the 0.45μm pores [65-66]. Therefore, a $MFI_{0.05}$ using membranes with pores of 0.05μm was developed. However, it is likely that particles smaller than 0.05μm which are not retained by the test membranes, are responsible for the observed flux decline rates [65-66].

Table 1.2:   Modified Fouling Index (MFI) measured for pretreated River Rhine Water using membranes of different pore sizes [65].

Membrane Pore Diameter (μm)	Modified Fouling Index (s/l²)
0.8	4
0.4	60
0.2	200
0.1	1800
0.05	4500
0.015	2 000 000

From literature the presence of small particles is known in river and lake water. Particle size distribution measurements for River Rhine and lake water in Switzerland and for the Mississippi in the USA, showed that although, smaller particles 50-300nm only contributed a small percentage (≈2%) of the total particle mass they were present in the greatest number [51,68-70]. Moreover, the size distribution of smaller particles may vary seasonally. In one Swedish lake water study, smallest colloids were found in spring (range 120-340) while in autumn and summer the size distribution was in the range of 280-70nm [51]. Schippers showed that MFI is inversely dependent on particle size through the Carmen-Kozeny equation for specific cake resistance [66]. Therefore, the fouling potential of smaller particles is higher and not measured in the existing MFI ($MFI_{0.45}$ and $MFI_{0.05}$) tests. Consequently, until now, there is no method available to measure and assess the fouling potential of these particles in a feedwater and over time. Furthermore, the efficiency of various pretreatment processes *e.g.* slow sand filtration in removing smaller colloidal particles and in reducing the particulate fouling potential of a feedwater can not be evaluated and monitored. Therefore, a new parameter needs to be developed to include the fouling potential of these smaller particles.

1.4   Goal and Objectives of the Research

It is widely recognised that the control of scaling and particulate fouling is instrumental in further membrane technology advancement and in decreasing costs associated with membrane filtration applications not only in drinking and industrial water treatment but in waste water reuse and recycling. However, predictive models for scaling and particulate fouling have not yet proven successful for general use, in part their unreliability is due to the uniqueness of each combination of feedwater composition, membrane type and pretreatment schedule. Therefore, pilot plant operation is commonly used for the prediction of scaling and particulate fouling to design full scale systems as this method generally provides reasonably good reproducibility.

The control of scaling and particulate fouling in membrane filtration systems can be brought about by a clearer understanding of the processes involved in these phenomena and more accurate methods to predict and prevent these phenomena, which is the goal of this research. Specific research objectives are to establish (1) the solubility and kinetics of scaling and to develop an approach for scaling prediction, using barium sulphate as a model scalant and (2) an accurate predictive test to determine the particulate fouling potential of a feed water (further development of the Modified Fouling Index making use of ultrafiltration membranes). These tests can be used as tools to determine the efficiency of particulate fouling and scaling prevention techniques, for improvements thereof in the absence of expensive pilot plant studies.

References

1.   J.C. Schippers, Desalination methods, IHE Lecture notes.
2.   S. Postel, State of the World 2000, L. R. Brown, C. Flavni and H. French (Eds.), W.W. Norton and Company, Inc., New York, 2000 41-47
3.   W. Owens and K. Brunsdale, Solving the problem of fresh water scarcity in Israel, Jordan, Gaza and the West Bank, Water for Peace in the Middle East and Southern Africa, Green Cross International (2000) 67-74.
4.   S. Pohoryles, Program for efficient water use in middle east agriculture, Water for Peace in the Middle East and Southern Africa, Green Cross International (2000), 18-32.
5.   M. A. Al-Sahlawi, Seawater desalination in Saudi Arabia: economic review and demand projections, Desal., 123, (1999), 143-147.
6.   K. Wangnick, IDA Worldwide desalting plants inventory report, No. 16, Wangnick consulting, GMBH, (2000).
7.   J.C. Schippers and J. C. Kruithof, Membraanfiltratie over 10 tot 25 jaar, H_2O, 6, 1997, 179-182.
8.   C. Anselme, V. Mandra, I. Baudi and J. Mallevialle, Optimum use of membrane processes in drinking water teatment, IWSA conference Madrid, SS 2-1-SS2-11.
9.   R. Rautenbach, K. Vossenkaul, T. Linn, and T. Katz, Waste water treatment by membrane processes - New developments in ultrafiltration, nanofiltration and reverse osmosis, Desal., 108 (1-3) (1996) 247-253.
10.   F.P. Cuperus and C.A. Smolders, Characterization of UF membranes. Membrane characteristics and characterization techniques, Adv. in Coll. Inter. Sci, 34, (1991) 135-173.
11.   T.M. Dawes, G.L. Leslie, W.R. Mills, J.C. Kennedy, D.F. McIntyre, B.P.Anderson,The groundwater replenishment system, Desal and Water Reuse, 10/2 (2000), 55-61.
12.   C. Ventresque, V. Gisclon, G. Bablon and G. Chagneau, An outstanding feat of modern technology: the Mery-sur-Oise Nanofiltration Treatment Plant (340,000 m^3/d) Desal, 131, (2000) 1-16.
13.   P.C. Kamp, J.C. Kruithof and H.C. Folmer, UF/RO treatment plant Heemskerk: from challenge to full scale application, Desal, 131, (2000), 27-35.
14.   O.A. Hamed, A.G.I. Dalvi, M.N.M. Kither, G.M. Mustafa, K. Bamardouf,M.A. Al-Sofi and A.M. Hassan, Optimization of hybridized seawater desalination process, Desal, 131, (2000), 147-156.
15.   A.M. Hassan, A.M. Farooque, A.T.M. Jamaluddin, A.S. Al-Amoudi, M.A.Al-Sofi, A.F. Al-Rubaian, N.M. Kither, I.A.R. Al-Tisan and A. Rowaili, A demonstration plant based on the new NF-SWRO process, Desal, 131, (2000), 157-171.

16.   J.C. Schippers, Presentation Membranes in Drinking and Industrial Water Production, Paris, France, 3-6 October 2000.

17.   O.A. Hamed, M.A.K. Al-Sofi, G. M. Mustafa and A. G. Dalvi, The performance of different anti-scalants in multi-stage flash distillers, Desal. 123 (2-3) (1999), 185-194.

18.   S. Patel and M. A. Finan, New antifoulants for deposit control in MSF and MED plants, Desal. 124 (1-3) (1999), 63-74.

19.   E.G. Darton, RO plant experiences with high silica waters in the Canary Islands, Desal. 124 (1-3) (1999), 33-41.

20.   I. Bremere, M.D. Kennedy, S. Mhyio, A. Jaljuli, G. Witkamp and J.C. Schippers, Prevention of silica scale in membrane systems: removal of monomer and polymer silica, Desal., (2000) 132, 89-100.

21.   Handbook of Chemistry and Physics, 1978.

22.   J.W.Mullin, Crystallization, 3rd Edition. Butterworth-Heinemann, 1993.

23   O.Söhnel and J.Garside, Precipitation - Basic Principles and Industrial Applications, Butterworth-Heinemann, 1992.

24.   C.C.Templeton, Solubility of barium sulphate in sodium chloride from 25° to 95°C, J. of Chem. and Eng. Data, 5 (1960) 514-516.

25.   S. Lee, J. Kim and C.H. Lee, Analysis of $CaSO_4$ scale formation mechanism in various nanofiltration modules, J.of Mem. Sci. 163 (1) (1999), 63-74.

26.   L.Y. Dudley and E.G. Darton, Pretreatment procedures to control biogrowth and scale formation in membrane systems, Desal. 110 (1-2) (1997), 11-20.

27.   Permasep Engineering Manual (PEM). Du Pont de Nemours & Co. 1982.

28.   F.H. Butt, F. Rahman and U. Baduruthamal, Characterization of foulants by autopsy of RO desalination membranes, Desal. 114 (1) (1997), 51-64.

29.   J.S. Baker, S.J. Judd and S.A. Parsons, Antiscale magnetic pretreatment of reverse osmosis feedwater, Desal., 110 (1-2) (1997), 151-165.

30.   K.W. Busch and M.A. Busch, Laboratory studies on magnetic water treatment and their relationship to a possible mechanism for scale reduction, Desal. 109 (2) (1997) 131-148.

31.   I. Bremere, M.D. Kennedy, A. Johnson, R. van Emmerik, G. Witkamp,and J.C. Schippers, Increasing conversion in membrane filtration systems using a desupersaturation unit to prevent scaling, Desal. 119 (1-3) (1998), 199-204.

32.   Ingrida Bremere, Maria Kennedy, Peter Michel, Rani van Emmerik, Geert-Jan Witkamp and Jan Schippers, Controlling scaling in membrane filtration systems using a desupersaturation unit, Desal. 124 (1-3) (1999), 51-62.

33.   M.C. Van der Leeden, The Role of polyelectrolytes in barium sulfate   Precipitation. Ph.D Thesis, TU Delft, 1991.

34.   A. Al-Ahmad and F.A. Aleem, Scale formation and fouling problems and their predicted reflection on the performance of desalination plants in Saudi Arabia. Desal., 96, (1994) 409-419.

35.   J.A.M. Paassen, J.C. Kruithof, S.M. Bakker and F. Kegel-Schoonenberg, Integrated multi-objective membrane systems for surface water treatment: pre-treatment of nanofiltration by riverbank filtration and conventional ground water treatment, Desal., 118 (1998), 239-248.

36.   J.P. Van der Hoek, P.A.C.Bonné and E.A.M. Van Soest, Fouling of Reverse Osmosis Membranes: The effect of pretreatment and operating conditions. Proceedings of the American Waterworks Association 1997 Membrane Technology Conference, February 23-26, New Orleans, U.S.A. 1029-1041.

37.   E.G. Darton, Membrane chemical research: centuries apart, Desal. (2000), 161-171.

38.   J. S. Gill, A novel inhibitor for scale control in water desalination, Desal. 124 (1-3) (1999), 43-50.

39.   A.E. Jaffer. The application of novel chemical treatment program to mitigate scaling and fouling in RO Units. Desal., 96 (1994) 71-79.

40.   A.L. Rohl, D.H. Gay, R.J. Davey and C.R.A. Catlow, Interactions at the organic/inorganic interface: molecular modelling of the interaction between diphosphonates and the surfaces of bartie crystals, J. Am. Soc, (1996), 118 642-648.

41.   J.C.Cowan, Water-Formed Scale Deposits, Gulf Publishing Company, 1992.

42.   A. G. Pervov, Scale formation prognosis and cleaning procedure schedules in RO system operation, Desal., 83 (1991) 77-118.

43. R. Sheikholeslami and S. Tan, Effects of water quality on silica fouling of desalination plants, Desal. 126 (1-3) (1999), 267-280.

44. A. Quddus and I.M. Allam, $BaSO_4$ scale deposition on stainless steel, Desal. 127 (3) (2000), 219-224.

45. K.S. Pitzer, Activity Coefficients in Electrolyte Solutions, 2nd Edition. CRC Press, 1991.

46. J.Buffle, Complexation Reactions in Aquatic Systems. Ellis Horwood Limited 1988.

47. S. R. Ahmed, M. S. Alansari and T. Kannari, (1989) Biological fouling and control at Ras Abu Jarjur RO plant - a new approach. Desal., 73, 69-84.

48. G. Peplov and F. Vernon, Trace metal fouling and cleaning of seawater RO membranes. Desal., 66, (1987), 271-284.

49. R. Ning, Reverse osmosis process chemistry relevant to the Gulf. Desal., 123, (199), 157-164

50. D. Perret, M.E. Newman, J.C. Negre, Y. Chen, and J. Buffle, Submicron Particles in the Rhine River - I. Physico-Chemical Characterisation, Wat. Res. 28,1 (1994), 91-106.

51. A. Ledin, S. Karlsson, A. Duker and B. Allard, Characterization of the submicrometre phase in surface waters, Anal. 120, (1995), 603-608.

52. D. Van der Meent, A. Los, J.W. Leeuw and P.A. Schenck, Size fractionation and Analytical pryloysis of suspended particles from the Rver Rhine delta, Adv. in Org. Geochem., (1981) 336-349.

53. H. Peavy, D.R. Rowe and G. Tchobanoglous, Environemental Engineering, McGraw Hill.

54. A. G. Fane, K. J.Kim, P.H. Hodgson, G. L.eslie, C.J.D. Fell, A.C.M. Franken, V. Chen, and K.H. Liew, (1992) Strategies to Minimise Fouling in the Membrane Processing of Biofluids, Amer. Chem. Soc., 2, 92, 304-319.

55. J. Hermia, Étude analytique des lois de filtration à pression constante, Rev. Univ. Mines, 2 (1966) 45-51.

56 M. Kennedy, S. Siriphannon, I. .van Hoof and J. Schippers, Improving the performance of dead-end ultrafiltration systems: comparing air and flushing, Conference Membranes in Drinking and Industrial Water Production, Paris, France, 3-6 October (2000) 1, 373-382.

57. M. Kennedy, S. Kim, I. Mutenyo, L. Broens and J. Schippers, Intermittent crossflushing of hollow fiber ultrafiltration systems, Desal. 118 (1-3) (1998), 175-187.

58 G. Galjaard, J. van Paassen, P. Buijs and F. Schoonenberg, Enhanced pre-coat engineering (EPCE) for micro- and ultrafiltration: the solution for fouling?, Conference Membranes in Drinking and Industrial Water Production, Paris, France, 3-6 October (2000) 1, 605-610.

59. J.G. Jacangelo, J.M. Laîné, K.E. Carns, E.W. Cummings and J. Mallevialle, Low pressure membrane filtration for removing Giardia and microbial indicators, J. AWWA, (1991),83:9:97-106.

60. S.S. Adham, J.G. Jacangelo and J.M. Laîné, Low pressure membranes: assessing integrity, J. AWWA, (1995),87:3: 62-75.

61. P. Lipp, G. Baldauf, R. Schick, K. Elsenhans and H-H. Stabel, Integration of ultrafiltration to conventional drinking water treatment for a better particle removal-efficiency and costs, Desal., 119 (1998), pp.131-142.

62. J.C Kruithof, J.C. Schippers, P.C. Kamp, HC Folmer and J.A.M.H. Hofman, Integrated multi-objective membrane systems for surface water treatment: pretreatment of reverse osmosis by conventional treatment and ultrafiltration, Desal., 117 (1998) 37-48.

63. K.E. Morris and J.S. Taylor, The use of Fouling Indices to predict reverse osmosis membrane Fouling. 1991 AWWA Proceedings. Membrane Technologies in the Water Industry. Orlando 587-599.

64. J.C. Schippers and J. Verdouw, The Modified Fouling Index, a method of determining the fouling characteristics of water, Desal., 32, (1980), 137- 148.

65. J.C. Schippers, J.H. Hanemaayer, C.A. Smolders and A. Kostense, Predicting flux decline of reverse osmosis membranes, Desal., 38, (1981), 339-348.

66. J.C. Schippers, 1989 PhD Thesis ISBN 90-9003055-7, Vervuiling van hyperfiltratiemembranen en verstopping van infiltratieputten, Keurings instituut voor Waterleiding-artikelen Kiwa N.V, Rijswijk, The Netherlands.

67. J.C. Schippers and A. Kostense, The effect of pretreatment of River Rhine water on fouling of spiral wound reverse osmosis membranes, Proc., of the 7th International Symposium on Fresh Water from the Sea, (1980) Vol. 2 297-306.

68. M.E. Newman, M. Fiella, Y. Chen, J.C. Negre, D. Perret,and J. Buffle, Submicron particles in the Rhine River - II. Comparison of field observations and model predictions, Wat. Res. 28,1 (1994), 107-118.

69.   M. Fiella, J. Buffle and G.G. Leppard, Characterization of submicron colloids in freshwaters: Evidence for their bridging by organic structures, Wat. Sci. Tech., 27, 11 (1993), 91-102.
70.   T.F. Rees and J. F. Ranville, Collection and analysis of colloidal particles transported in the Mississippi River, USA, J. Contaminants Hydol., 6, (1990) 241-250.

2

Barium Sulphate Solubility Prediction in Reverse Osmosis Membrane Systems

Chapter 2 is based on: "BaSO₄ Solubility in Reverse Osmosis Concentrates" by S.F.E. Boerlage, M.D. Kennedy, G.J. Witkamp, J.P. Van der Hoek, J.C. Schippers. Published in *Journal of Membrane Science*, Vol. 159 (1999), pp. 47-59, and "Prediction of BaSO₄ Solubility in Reverse Osmosis Systems" by S.F.E. Boerlage, M.D. Kennedy, G.J. Witkamp, J.P. Van der Hoek, J.C. Schippers. Proceedings of the American Waterworks Association (1997) Membrane Technology Conference, February 23-26, New Orleans, U.S.A., pp. 745-759.

Contents

Abstract

Barium sulphate scaling in reverse osmosis (RO) causes flux decline and potentially severe membrane damage. The method in the Du Pont Manual to predict barium sulphate scale, based on predicting barium solubility in RO concentrate, is unreliable and limited to 25°C. This method predicted barium solubility was exceeded 14 times at 80% recovery and yet no scaling occurred at the pilot plant. Possible explanations are; inaccurate solubility prediction, low rate of barium sulphate precipitation and/or organic matter effects on solubility or precipitation. This study investigated barium solubility in the RO concentrate and the effect of *e.g.* ionic strength using more theoretical approaches to solubility prediction *i.e.* Bromley and Pitzer models. Seeded growth determination of barium solubility in RO and synthetic concentrate (no organic matter) confirmed supersaturation and proved organics had no effect on solubility. Du Pont's method under predicted solubility by *circa* 30% at 25°C. The Pitzer and Bromley methods when calibrated for RO concentrate by applying an experimental K_{sp} gave accurate prediction at 5-25°C for the ionic strength range of 0.01-0.1M. For higher ionic strengths, the Pitzer model was more accurate. The observed stable supersaturation (27 times the solubility at 5°C) in the pilot plant is most likely due to the low rate of barium sulphate precipitation.

2.1   Introduction

Traditionally, reverse osmosis (RO) was applied for the desalination of brackish water and seawater. Nowadays, RO is increasingly being considered for the treatment of surface water sources. Reverse Osmosis is under investigation at Amsterdam Water Supply (AWS) as an integrated part of two treatment systems. In both systems, River Rhine water is pretreated by coagulation, sedimentation, and rapid sand filtration. Subsequently, in scheme I, the pretreated water is treated by ozonation - biologically activated carbon filtration - slow sand filtration and reverse osmosis (RO I). Whereas, scheme II comprises slow sand filtration prior to reverse osmosis (RO II) [1]. In these schemes RO is applied to achieve multiple objectives including the removal of salts, hardness, pesticides and other organic micropollutants, and to ensure adequate disinfection.

A major problem facing RO is membrane scaling. Scaling refers to the precipitation of sparingly soluble salts on RO membranes. This phenomenon may lead to flux decline or increased feed pressure and eventually serious membrane damage. In particular, barium sulfate ($BaSO_4$) present in RO feedwater may lead to scaling problems due to its low solubility, (1×10^{-5} mol/L in pure water) [2]. $BaSO_4$ scale is problematic as the early stages are difficult to detect, due to the low sensitivity of the currently applied pressure or flow measurement systems. If the scale layer is detected at a later stage when it has aged into a hard adherent layer it may be resistant to conventional RO cleaning methods. It can then only be dissolved by crown ethers and concentrated sulfuric acid, which will cause membrane hydrolysis [2].

Scaling can be avoided by lowering the recovery below the solubility of the scalant ion. However, high recoveries are economically desirable since water production is maximized and concentrate production is minimized. To maintain recoveries and prevent scaling, acids and antiscalants are often applied. Antiscalants are expensive and often are biodegradable, potentially leading to membrane biofouling and organic fouling [3]. This is because antiscalants are designed to have a certain degree of biodegradability in order to avoid environmental pollution arising from concentrate disposal. However, the biodegradability can have an adverse effect when it occurs within the membrane system [3].

Accurate scaling prediction is of great importance in every RO plant, dictating the design recovery which can be achieved and whether antiscalant addition is required to prevent scaling. The method in the Du Pont Manual [4] is frequently used for scaling prediction in the RO industry. The method predicts whether the barium sulphate solubility will be exceeded (*i.e* solution is supersaturated) at the desired recovery.

This work focuses on the AWS pilot plant, where although, the concentration of barium in the feedwater is low (40-90μg/L) the sulphate concentration is high (40-80mg/L) and according to the Du Pont method the predicted solubility of the feedwater is exceeded 1.5 times. Thus, the feedwater is already supersaturated and scaling is expected to occur. However, the system has been operated at 80% recovery without an antiscalant and no scaling occurred during the one year of operation at this recovery, despite the solubility being predicted to be exceeded 14 times. This "stable" supersaturation phenomenon illustrates the limitations of the Du Pont method in predicting the recovery at which scaling occurs in RO systems. Moreover, in AWS seasonal temperature variations can range from 25°C in summer to close to 0°C in winter while the Du Pont method is limited to 25°C, and from literature barium solubility is expected to decrease at lower temperatures [5]. Therefore, the supersaturation and thus the driving force for scaling could be even higher at lower temperatures.

The difference between predicted and actual scale encountered may be attributable to the following hypotheses:

(i).   inaccurate solubility prediction *i.e.* concentrate are not supersaturated and/or
(ii).   the presence of organic matter in surface waters *e.g.* humic and fluvic acids are known to form complexes with metal ions such as barium and calcium [6]. Complexation of barium by organic matter in the RO concentrate would increase barium solubility and reduce supersaturation.
(iii).   if the concentrate are indeed supersaturated, stable supersaturation known in literature as *metastability,* may occur due to the low rate of precipitation kinetics [7].

This paper focuses on the first two hypotheses in a solubility investigation to determine if the AWS feedwater and concentrate are supersaturated and to develop an accurate solubility prediction method in RO concentrate for the temperature range of 5-25°C. Three different approaches, namely Du Pont, Bromley (extension of Debye Hückel) and Pitzer, are evaluated. In addition, the role of organic matter on barium solubility is investigated by estimating the potential amount of barium that could be complexed and by a comparison of barium solubility in synthetic (no organic matter present) and RO surface water concentrate.

2.2   Background

The *concentration solubility product* or simply the *solubility product* (K_C) for $BaSO_4$, is [7,8]:

$$K_C = [Ba^{2+}][SO_4^{2-}] \quad \text{(at equilibrium)} \qquad ...(2.1)$$

$[Ba^{2+}]$ and $[SO_4^{2-}]$ are expressed as concentration in molarity (mol/L). If either barium or sulphate is added to the saturated solution through another source *e.g.* sulphuric acid addition, then the supersaturation of the solution will increase and the overall solubility will decrease through the *common ion effect* in accordance with the K_C and equilibria principles. K_C is not constant at a defined pressure and temperature but is dependent on ionic strength, *I*:

$$I = \frac{1}{2} \sum M_i z_i^2 \qquad ...(2.2)$$

Where M_i is molal concentration (mol/L water) and z_i is ionic charge of ion *i*. At ionic strengths > 0.001M, electrical interactions occur between ions, resulting in an increase in solubility through the *ionic strength effect* [7,9]. Thus, using molarities in the K_C calculations, is no longer accurate and K_C should be corrected for ionic strength. Alternatively, activities should be used which take into account ionic interactions, and are expressed in the *thermodynamic solubility product* K_{sp} [8]:

$$K_{sp} = a_{Ba^{2+}(aq)} \, a_{SO_4^{2-}(aq)} = \gamma_+ [\,Ba^{2+}\,]\gamma_- [\,SO_4^{2-}\,] \qquad ...(2.3)$$

where the activity $a_{i(aq)}$ is the product of ion i molal concentration and its *activity coefficient* γ_i, (dimensionless) which corrects for interionic forces occurring in solution. The thermodynamic solubility product (K_{sp}) is independent of ionic strength at a defined temperature and pressure.

2.2.1   Du Pont

The Du Pont method predicts barium solubility at 25°C using a *solubility product*, K_C, in which the K_C is read graphically as a function of ionic strength for the desired recovery. The relationship between K_C and I was derived as an average from data on the effect of individual monovalent and divalent cations on barium solubility from Davis [10]. The K_C is then compared with the product of the scalant ions (molarity). When the scalant concentration product exceeds K_C, the solution is supersaturated and scaling can occur. To avoid precipitation, Du Pont [4] recommends a recovery such that the scalant concentration product is 20% below the K_C, thus ensuring a safety factor for concentration polarisation which may occur at the membrane surface which causes a higher localized concentration than in the bulk solution:

$$\left[Ba^{2+}\right]\left[SO_4^{2-}\right] = 0.8\,K_C \qquad ...(2.4)$$

2.2.2   Bromley

The Debye Hückel theory was derived from a simplified model of an electrolyte solution using statistical mechanics to derive theoretical expressions for activity coefficients of ions in solution. Only very dilute solutions were considered where the main deviation from ideality was assumed to be due to long-range Coulomb interactions between ions. The final result is the *Debye-Hückel Limiting Law*, Equation 2.5. Debye Hückel is generally accurate for solutions with an ionic strength ≤ 0.01M. [11].

$$\log \gamma_{\pm} = -\left| z_{Ba^{2+}} \, z_{SO_4^{2-}} \right| A_o' \sqrt{I} \qquad ...(2.5)$$

where $\gamma_{\pm}$ is the mean activity coefficient, A_o' is 0.509 $(mol/L)^{-0.5}$ for aqueous solutions at 25°C. Activity coefficients in solutions up to 6M can be estimated from the Bromley correlation (25°C) an extension of the Debye Hückel Limiting Law which employs a constant B_1 (-0.037 for $BaSO_4$) for ion interaction [12].

$$\frac{1}{z_+ z_-} \log \gamma_{\pm} = -0.511\frac{\sqrt{I}}{1+\sqrt{I}} + \frac{(0.06+0.6B_1)\,I}{1+\sqrt{I}} + \frac{B_1\,I}{z_+ z_-} \qquad ...(2.6)$$

Where the B_1 constant is composed of ionic interactions for barium and sulphate.

$$B_1 = B_+ + B_- + \delta_+ \delta_- \qquad \qquad ...(2.7)$$

2.2.3   Pitzer

The Pitzer specific ion interaction model [13] expands upon the Debye Hückel theory by taking into account *long and short range* interactions between ions in solution in calculating activity coefficients. The Pitzer model is valid for ionic strengths up to 6M [14]. Pitzer developed a set of equations which describe the properties of electrolyte solutions by an electrostatic term for the long range interactions and a virial coefficient series to account for the short range specific ion interactions. The simplified form of the Pitzer equations are expressed as: [15]:

$$\ln \gamma_M = z_M^2 F_1 + \sum_a c_a (2 B_{Ma} + Z C_{Ma}) +$$
$$+ z_M \sum_c \sum_a c_c c_a C_{ca} \qquad \qquad ...(2.8)$$

$$\ln \gamma_X = z_X^2 F_1 + \sum_c c_c (2 B_{cX} + Z C_{cX}) +$$
$$+ |z_X| \sum_c \sum_a c_c c_a C_{ca} \qquad \qquad ...(2.9)$$

where γ_M and γ_X are the activity coefficients of ions of interest, subscripts a, X and c, M denote anion, anion of interest, cation and cation of interest, respectively. The first term in the equations, F_1, arises from the long range electrostatic interactions between $_M$ and $_X$ ions through an improved analysis of the Debye Hückel Limiting Law. The following terms arise from the second virial coefficient B which describes the ionic strength dependence of short range interactions between $_M$ and $_X$ ions. The third virial coefficient C is important at high concentrations and for triple ion interactions. Z is a constant dependent on ionic charge and concentration. The equations for F_1, B, C and Z can be found in Pitzer [15].

2.3   Materials and Methods

2.3.1   Determination of Barium Solubility

2.3.1.1   Materials

Analytical reagent grade chemicals were used, deionised water, pretreated plastic bottles and Grade A glassware (preparation of standards). Labware was pretreated by soaking in 7M HNO$_3$ for 24 hours, and rinsing three times with deionised water, prior to use. Barium standards, 0-150µg/L, and 0-1000µg/L for feedwater and concentrate, respectively, were prepared from a solution of 1000µg/L barium (Baker) using calibrated Eppendorf pipettes, concentrated HNO$_3$

(INSTRA-Analyzed, Baker) and deionised water. A certified standard solution of 100µg/L barium (Perkin Elmer), was used to check the accuracy of calibration. Seed crystals of commercially available $BaSO_4$ (Baker Analysed) were used after aging for 24 hours, in 1mL deionised water at 25°C.

2.3.1.2   Synthetic Water

Deionized water (8L) was stored at 25°C (or 5°C) for 12-15 hours preceding salt addition to allow temperature equilibration, in order to prevent any artefacts occurring in barium solubility due to a temperature change from the desired experimental temperature. All salts were analytical grade and were added to the temperature equilibrated deionized water to reach the same chemical composition and pH of AWS concentrate. The sequence of addition of the required amounts of salts was as follows; One litre containing the desired concentration of $NaCl$, KNO_3 and $MgSO_4$, was added to one litre of water. The pH of the solution was measured. After which, a further one litre of water containing the desired concentration of $NaHCO_3$ and Na_2SO_4 was added. The pH was remeasured and adjusted to 8 with concentrated HCl (and 10% NaOH) to 7.4 to prevent precipitation of $CaCO_3$ when $CaCl_2$ is added. Three more litres of water were added followed by one litre of water containing the desired concentration of dissolved $CaCl_2$. The pH was maintained at 7. The final litre of water added, contained the required concentration of barium (prepared by dilution of barium standard 1000µg/L). The pH was adjusted to 6.8-7.1.

The anion concentration in the synthetic water was checked by ion chromatography (Dionex Series 4500i with integrator, Shimatzu C-R5A Chromatopac). Cations were analysed by inductively coupled plasma spectrophotometry. The concentration of both cations and anions were within 5% of the desired concentration. The background concentration of barium in the deionized water was below the detection limit of inductively coupled plasma spectrophotometry i.e. 0.2µg/L.

2.3.1.3   Barium Analysis

Barium was analysed by Inductively Coupled Plasma Spectrophotometry (ICP) (Spectroflame 21/016) and calibrated using the prepared barium standards. The calibration line comprised the blank, and barium standards ranging from 20-1000µg/L. The accuracy of calibration was checked with the certified barium standard of 100µg/L, the calibration was unacceptable if the standard deviation of the certified standard was greater than 5%.

2.3.1.4   Seeded Growth Technique

The seeded growth technique was employed, where the relative supersaturation of RO and synthetic feedwater and concentrate were disturbed by adding "seed" crystals. The "free-drift" method was applied, in which seed crystals either dissolve and form a saturated $BaSO_4$ solution or grow by depletion of $BaSO_4$ from the supersaturated bulk solution. The resultant increase or decrease in barium concentration after addition of seed crystals, due to dissolution or growth of the seed crystals, respectively, was detected using ICP. When no change in the barium concentration over time was found after *circa.* 3-24 hours, the equilibrium barium solubility was reached.

The seed crystal dose (in suspension) was added to batch reactors containing RO and synthetic feedwaters and concentrates (450mL), pre-equilibrated to 25°C or 5°C, and shaken on a Rotator-Model G2 at 200rpm for the experiment duration at 25°C or 5°C. The corresponding blanks (without seed crystal addition) were treated identically. Experiments were carried out in duplicate. Periodically 10mL samples in duplicate were taken from the batch reactors using a plastic syringe (Becton Dickinson) with a 0.2μm filter (Sartorius) rinsed with 0.2 M HNO_3, prior to use. The first 2mL of sample filtered was discarded and the following 8mL of filtrate collected, 1 drop of concentrated (14 M) HNO_3 was added to each sample prior to ICP analysis. The accuracy of the determination of the barium analysis of four samples is ±10% deviation of the average.

2.3.2   Prediction of Solubility

2.3.2.1   RO Feedwater and Concentrate Input Data

Data for the prediction of barium solubility in the RO feedwater and concentrate is given in Table 1. HCl is added to the feedwater to give a Langelier Saturation Index value of 0 to prevent alkaline scaling. The composition of the acidified pilot plant feedwater (from Scheme I and II) was determined three times and averaged. Whereas, for the scalant ions, the average concentration of barium and sulphate measured during the solubility investigation, 70μg/L and 58mg/L, respectively was used. The concentration of sodium and chloride was adjusted to give charge neutrality. The only difference between the two schemes is in the dissolved organic carbon concentration. The ozonation and activated carbon filtration pretreatment step in RO I breaks down and reduces the concentration of organic matter. The concentrations of all ions were multiplied by the corresponding concentration factor (CF), assuming 100% membrane rejection, at the desired recovery (Y) to obtain the concentrate ion balance:

$$CF = \frac{1}{1 - Y} \qquad \qquad ...(2.10)$$

Table 2.1:   Acidified AWS feedwater ion balance with average barium and sulphate concentrations after charge balance

Ion	RO Feedwater (mg/L)	After Charge Balance (mmol/L)
Hydrocarbonate	86	1.45
Chloride	142	5.02
Sulphate	58	0.6
Nitrate	18	0.29
Potassium	6	0.16
Magnesium	11	0.46
Calcium	70	1.75
Barium	0.07	0.0005
Sodium	78	3.39
pH	6.0-6.5	6.8-7.0
Dissolved Organic Carbon	1 (RO I)	1
	2 (RO II)	2
Total Dissolved Solids	470 (RO I)	506 (RO I)
(mg/L)	471 (RO II)	507 (RO II)

2.3.2.2   Du Pont Solubility Prediction

The K_C value was read from the Du Pont graph (refer Figure 2.1) at the calculated ionic strength for the desired recovery. The safety factor was neglected in this research as the concentration polarisation factor is minimal (estimated to be only 1.01 times the bulk concentration). The predicted solubility of barium (µg/L), s, was determined from:

$$s = \frac{K_C}{c_{SO_4^{2-}}} M_{Ba^{2+}} \qquad \text{...(2.11)}$$

$M_{Ba^{2+}}$ is molecular weight of barium (g/mol) and c is concentration of sulphate (mol/L) of the concentrate.

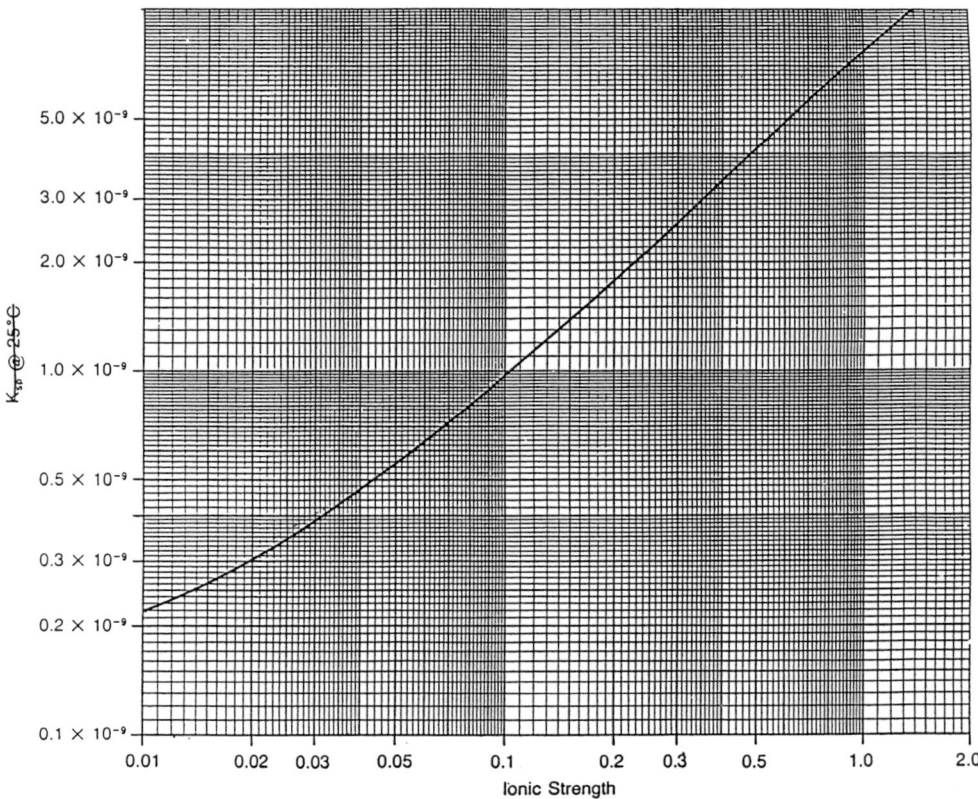

Figure 2.1: K_C for barium sulphate at 25°C versus ionic strength [4].

2.3.2.3   Pitzer and Bromley Solubility Prediction

A K_{sp} value (Equation 2.3) was estimated for Pitzer and Bromley solubility predictions by inputting the average *experimentally determined barium solubility* $C_{Ba^{2+}}$ (seeded growth experiments) and the average sulphate RO feedwater concentration and calculating the activity coefficients by the two approaches. For the calculation of activity coefficients in RO concentrate predictions, ion concentration and ionic strength were expressed as molarity. The difference between molarity (mol/L solution) and molality (mol/L water) at the highest ionic strength for the concentrates is 0.04%. However, for comparison of predictions with experimental solubility data in literature, molality was used

The BaSO$_4$ mean activity coefficient was calculated by Equation 2.6 for the Bromley Correlation at the calculated ionic strength of the RO concentrate. Barium solubility predictions were then determined using Equation 2.12 using the K_{sp} estimated using the Bromley approach.

$$s = \frac{K_{sp}}{c_{SO_4^{2-}} \, \gamma_{\pm}^2} M_{Ba^{2+}} \qquad\qquad \text{...(2.12)}$$

Separate activity coefficients for barium and sulfate ions, calculated for the Pitzer Equations (2.8 and 2.9) were computed using a computer program *Bluesea 1.0* [16]. Values for the second and third virial coefficients for single solutes *e.g.* NaCl, $NaHCO_3$ were obtained from Pitzer [17] and input into the program. The concentrate ion balance was input for each recovery and barium solubility was predicted using Equation 2.12, where $\gamma_{Ba^{2+}}$ the activity coefficient of barium replaces the mean activity coefficient and $a_{SO_4^{2-}}$ the activity of sulphate replaces the concentration of sulphate. The K_{sp} in Equation 2.12 refers to the K_{sp} estimated using the Pitzer approach.

For low temperature predictions when Pitzer coefficients were employed for temperature correction, only a minor effect (4%) was found on the activity coefficients between 0-30°C. As a result, the Pitzer coefficients at 25°C were used also for 5°C.

In the study investigating the role of organic matter on barium solubility the K_{sp} calculated using the Pitzer model was used with Equation 2.13 to express supersaturation as supersaturation ratio, S_r.

$$S_r = \sqrt{\frac{\gamma_+ [Ba^{2+}] \gamma_- [SO_4^{2-}]}{K_{sp_T}}} \qquad \qquad ...(2.13)$$

2.3.3 Estimation of Barium Complexed by Organic Matter

Humic acids, which may complex with barium ions, can account for up to 50% of dissolved organic carbon (DOC) in surface water [18]. In literature barium and calcium show approximately equal interaction with humic material [18]. Therefore, in order to estimate the maximum potential barium that could be complexed by organic matter in the 90% concentrate (RO II Line), it was assumed that (i) at 90% recovery, all the 20mg/L DOC was present as humic material and that all binding sites on the humic molecules are active for barium and calcium (ii) no competition between barium and calcium ions occurred for complexation sites and the exchange capacity is shared equally and proportionally to the respective molar concentration ratio *i.e.* 3:10 000, respectively and (iii) the average proton exchange capacity for commercial humic acids was estimated at 4.93 mmol of divalent metal per g of humic acid from literature [19, 20].

2.4 Results and Discussion

2.4.1 Experimentally Determined Barium Solubility

In Figure 2.2, an example of the determination of barium solubility in the feedwater is presented. The barium concentration in the blank (without seed crystals), was 68 ± 3µg/L for the entire experiment duration. Whereas, the barium concentration in the feedwater with seed crystal addition showed a significant increase due to dissolution of the $BaSO_4$ seeds. Equilibrium (saturation) was reached after *ca.* 3 hours. The t-Test confirmed that there was no statistical difference, between any of the solubilities after 3 hours, and at experiment termination (140 hours). No significant effect on solubility was found with the direct use of dry seed crystals or with aging the seed crystals for 24 to 2640 hours. The results indicated that on that day the feedwater was undersaturated with respect to barium sulphate.

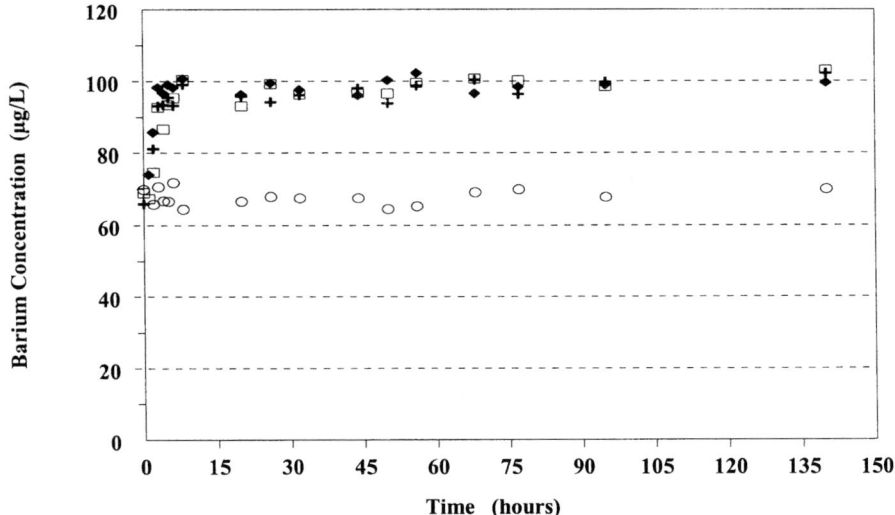

Figure 2.2:   Concentration of Barium in the RO Feedwater over Time at 25 °C. Seed Crystal Doses: 0mg (Blank) ○, 50mg ✛, 75mg □, and 300mg ◆.

Alternatively, the barium concentration decreased in the concentrate upon seeding with $BaSO_4$ crystals, due to growth of the seed crystals, indicating the concentrates were supersaturated.The barium solubility of pilot plant feedwater and concentrate determined at 25 °C are summarized as the lower line in Figure 2.3. A comparison of the lower line with the upper line, representing the actual barium concentration found in the concentrate in Figure 2.3 shows supersaturation. This confirms the Du Pont predictions that the concentrate are supersaturated with $BaSO_4$. The extent by which the concentrate are supersaturated (S_C) can be expressed as the ratio of the solution barium concentration c (mol/L) to the equilibrium barium solubility concentration c_{eq} (mol/L):

$$s_C = \frac{c}{c_{eq}} \qquad\qquad ...(2.14)$$

The average RO feedwater solubility was 81µg/L barium, which was 16% higher than the average barium concentration in the feedwater, indicating that in general the feedwater was slightly *undersaturated* (average S_C is 0.9). At increasing recoveries the supersaturation factor increased. The average S_C calculated for 50, 72, 80 and 90% recovery at 25 °C was 4.2, 6.5, 9.5 and 22.9, respectively. Thus, concomitantly the thermodynamic driving force for crystallization or scaling also increased.

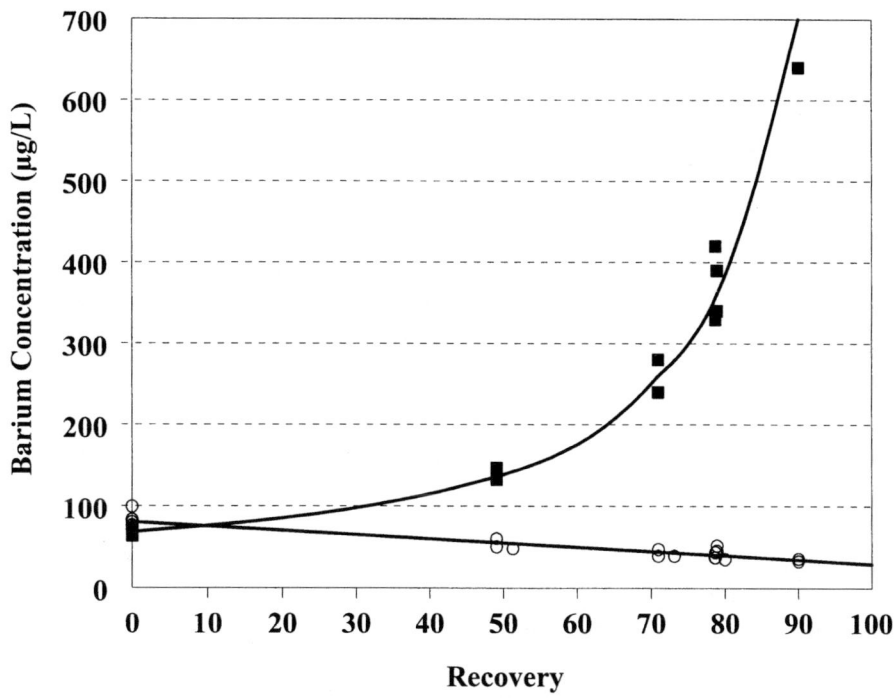

Figure 2.3: Experimentally determined barium solubility (○) as a function of recovery compared to measured barium concentration (■) of AWS plant feedwater and concentrate at 25°C.

2.4.2   Predicted Barium Solubility

The various solubility prediction methods were evaluated through a comparison of the barium solubility predicted and the average experimentally determined solubility. Barium solubilities predicted by the Du Pont method, the Bromley Correlation and the Pitzer model and the experimentally determined solubility for 25°C are presented graphically in Figure 2.4 and are given in Table 2.2.

Prediction of barium solubility using the Du Pont method is represented by the lower line in Figure 2.4. The Du Pont method predicted 30-40% lower solubilities than found experimentally. The Du Pont K_C varied from 2.25 - 10.5×10^{-10} for the ionic strength range of 0.01-0.1 molal for the feedwater and 90% concentrate, respectively. However, an ionic strength correction of the K_C alone is not enough for accurate prediction. The forces acting between ions in solution must be taken into account in solubility prediction through the use of activities.

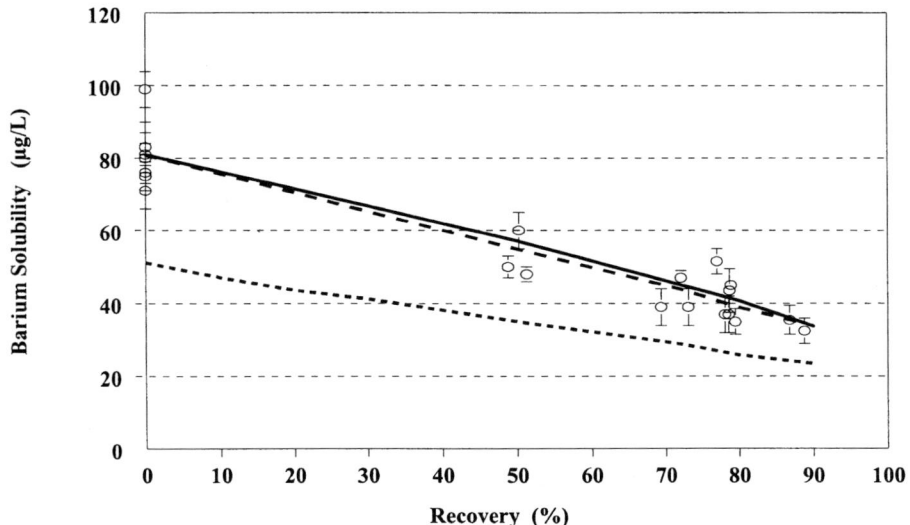

Figure 2.4: Experimentally determined barium solubility (○) of AWS feedwater and concentrate at 25 °C in comparison to solubility predicted by Du Pont (..) lower line, Bromley (--) middle line and Pitzer (-) upper line.

Table 2.2: Experimentally determined barium solubility in AWS feedwater and concentrate at 25 °C in comparison to solubility predicted by the method in the Du Pont Manual, Bromley Correlation and the Pitzer Model (with literature and experimental K_{sp}).

	Barium Solubility Predicted (μg/L) for RO feedwater and Concentrate		
	Feedwater	80%	90%
Du Pont	51	26	24
Bromley (Literature K_{sp})	59	29	25
Bromley (AWS K_{sp})	81	39	35
Pitzer (Literature K_{sp})	65	33	27
Pitzer (AWS K_{sp})	81	41	34
Average Experimental	81±8.1	42±4.2	34±3.4

Both the Bromley Correlation and Pitzer Model were calibrated to the AWS pilot plant, through the determination of the thermodynamic solubility product by seeding the RO feedwater. An experimental K_{sp} of 1.47×10^{-10} and 1.34×10^{-10} was found for the Bromley Correlation and Pitzer

Model, respectively. The difference in K_{sp} is due to the accuracy of the methods in calculating the activity coefficients. However, when the appropriate experimental K_{sp} was used in conjunction with the determination of activity coefficients for the prediction of barium solubility in the AWS concentrate very little difference was found between the barium solubility predicted (refer Table 2.2). Both approaches accurately predicted barium solubility in the AWS concentrates.

In the absence of an experimentally determined K_{sp}, the literature K_{sp} value for $BaSO_4$ in pure water of 1.08×10^{-10} [21] could be used for solubility prediction. This resulted in a lower solubility prediction of 26% for Bromley and 20% for Pitzer in comparison to the solubility predicted using the experimental K_{sp} (Table 2.2).

To prove the applicability of the Bromley Correlation and the Pitzer Model for general use for RO concentrate, another pilot plant installation (PWN Water Supply Company of North Holland) was investigated. The average concentration of sulphate was higher in the feedwater (136 mg/L) due to the addition of iron sulphate as a coagulant and the average barium concentration was 40μg/L. The feedwater was seeded to obtain the solubility to calculate the thermodynamic K_{sp}. The Bromley K_{sp} was calculated to be 1.37×10^{-10} and the Pitzer K_{sp} to be 1.17×10^{-10}. Results from both models again gave close solubility prediction to the experimental data when an experimentally determined K_{sp} was used (Table 2.3). Furthermore, in this case when the literature K_{sp} was used in the Pitzer model, accurate barium solubility prediction in the PWN concentrate was also found.

Table 2.3: Experimentally determined barium solubility in PWN feedwater and concentrate at 25°C in comparison to solubility predicted by Bromley Correlation and the Pitzer Model (with literature and experimental K_{sp})

	Barium Solubility Predicted (μg/L) for RO feedwater and Concentrate		
	Feedwater	63%	80%
Bromley (Literature K_{sp})	28	18	15
Bromley (PWN K_{sp})	36	23	20
Pitzer (Literature K_{sp})	32	21	18
Pitzer (PWN K_{sp})	34	23	19
Average Experimental	36±3.6	22±2.2	18±1.8

In summary, for the two pilot plants examined, more accurate prediction was obtained through the use of activities and an experimentally determined K_{sp} for both Bromley and Pitzer than by using a K_C corrected for ionic strength as in the Du Pont method. However, the Bromley Correlation is less accurate than the Pitzer Model when using only the K_{sp} pure water value. This is due to the fact that the forces acting between the ions in the concentrate are more accurately determined by the Pitzer Model, as the exact concentrate ion composition is input into the model in determining the activities. However, the Pitzer Model requires more time in inputting the concentration of each ion into the model for the desired recovery and inputting virial coefficients for all combinations of ions in solution.

2.4.3   Common Ion and Ionic Strength Effect on BaSO₄ Solubility

In the AWS and PWN feedwater the sulphate concentration is 1000 and 5000 times the barium concentration, respectively. Thus, the excess sulphate ion can act as a common ion which *decreases* barium solubility. The other relevant factor affecting solubility in RO concentrate is the increasing ionic strength with increasing recovery which *increases* barium solubility. These two factors will have opposing effects on barium solubility in the RO concentrate.

The effect of ionic strength on barium sulphate solubility in aqueous NaCl solutions was predicted using the Pitzer Model and Bromley Correlation for an eqimolar Ba^{2+} to SO_4^{2-} ratio and compared to experimental data from Templeton [22] and Davis [10] for 25 °C (Figure 2.5). The increase in barium sulphate solubility was more marked in dilute solutions, when the ionic strength was increased from 0.01 to 0.1 molal. This corresponds to feedwater and 90% recovery in RO surface water treatment.

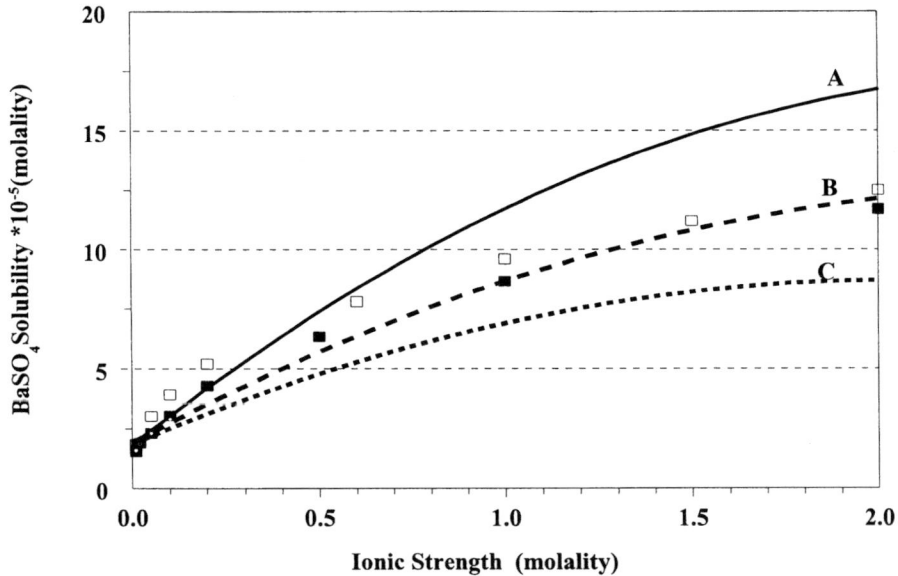

Figure 2.5: Experimental BaSO₄ solubility from Templeton □ [22] and Davis ■ [10] and solubility predicted by Bromley Correlation (line A) and Pitzer Model with optimised NaCl Pitzer coefficients (line B) and literature [17] NaCl coefficients (line C) using the literature K_{sp} value for pure water at 25 °C in aqueous NaCl Solutions of 0.01-2.0M ionic strength.

Predictions by Bromley (upper line in Figure 2.5) at ionic strength greater than 1 molal (the average ionic strength for seawater) were found to deviate from experimental values with the solubility still markedly increasing with ionic strength. Whereas, Pitzer predictions at the higher ionic strength were lower than solubility reported in literature but the diminishing increase in solubility with increasing ionic strength was found. The deviation of Pitzer model predictions from literature values at the higher ionic strength is due to the accuracy of virial coefficients for salts. Increasing the second virial coefficient for NaCl gave close solubility prediction (middle line in Figure 2.5) to experimental values in the 0.5-2.0 molal range.

However, in the RO concentrate an overall decrease in barium solubility is found despite the positive effect of increasing ionic strength at higher recoveries. This is in agreement with the "common ion effect", which occurs because of the higher sulphate to barium concentration. The barium solubility is 2.4 times lower at 0.1 molal than at 0.01 molal (refer Figure 2.3).

To quantify the individual effect of increasing ionic strength and sulphate concentration on barium solubility in the AWS concentrate, the Pitzer model was employed in two theoretical scenarios. The ionic strength was kept constant through the addition or subtraction of sodium and chloride ions while the sulphate was added according to the concentration factor for 90% recovery (CF=10). The ionic strength was maintained in Scenario (I) at 0.01 molar and in Scenario (II) at 0.1 molar. The results are given in Table 2.4.

Table 2.4: Barium Solubility (μg/L) predicted by the Pitzer Model for Scenario I (0.01 molar) and Scenario II (0.1 molar) with increasing sulphate concentration for the RO concentrate.

Ionic Strength (molar)	Sulphate Concentration		
	Feedwater 58 mg/L	90% Recovery 580 mg/L	Common ion effect factor of solubility decrease
Scenario I 0.01	81	13.6	6
Scenario II 0.10	316	34	9.3
Ionic Strength effect factor of solubility increase	3.9	2.5	-

The factor by which the barium solubility is predicted to increase due to the increasing ionic strength, decreases due to the higher sulphate concentration at 90% recovery from 3.9 to 2.5 *i.e.* less effective. A comparison of Scenario (I) to (II) shows that increasing the sulphate ion for Scenario (I) the solubility is predicted to decrease by a factor of *circa* 6 and for Scenario (II) by a factor of *circa* 9. In reality, the barium solubility decreased by 2.4. Thus, the increasing ionic strength is predicted to be responsible for a barium solubility increase in the AWS concentrate at 90% recovery of *circa* 20ug/L when the sulphate to barium stoichiometry is 1000:1. From the eqimolar Pitzer predictions in Figure 2.5 the barium solubility would increase by 1.80mg/L. This illustrates the potential of sulphate removal by *e.g.* anionic exchange in reducing the overall supersaturation and in increasing the solubility of barium in the concentrate and thus reducing the likelihood of scaling. However, regeneration of the ion exchange resin is cost prohibitive and therefore, this method would be of limited practical use.

2.4.4   Effect of Organic Matter on $BaSO_4$ Solubility

The results of the calculation estimated that maximum only *circa* 0.58% of barium and 0.56% of calcium present in the 90% concentrate could be complexed by humic material (Table 2.5). As the assumption was that all organic matter present could bind barium and there was no preference for calcium or barium the real amount is most likely far lower.

Table 2.5:   Maximum concentration of Barium and Calcium ions bound with humic acid at 90% recovery

Concentration of Barium Complexed		Concentration of Calcium Complexed	
2.96×10^{-8} mol/L	0.58%	9.86×10^{-5} mol/L	0.56%

As a further test to prove organic matter present in AWS concentrate has no effect on barium solubility, the solubility of the pilot plant, RO I and RO II, and synthetic (containing no organic matter) feedwater and concentrate were compared. The results obtained are presented in Figure 2.6 as a function of supersaturation ratio (using the Pitzer K_{sp}). From Figure 2.6 similar solubilities were found for synthetic and both RO I and RO II concentrate, of similar supersaturation. A difference may have been expected between RO I and RO II as RO I has 50% less organic matter and the organic matter has been broken down to smaller more polar molecules through the ozonation step.

Thus, it can be concluded that organic matter present in the AWS feedwater has a negligible effect on barium solubility. This indicates that the *metastability* of the concentrate is not due to the formation of barium complexes with organic matter which, if it occurred, would increase barium solubility and lower the supersaturation. However, organic matter may have an effect on the precipitation kinetics.

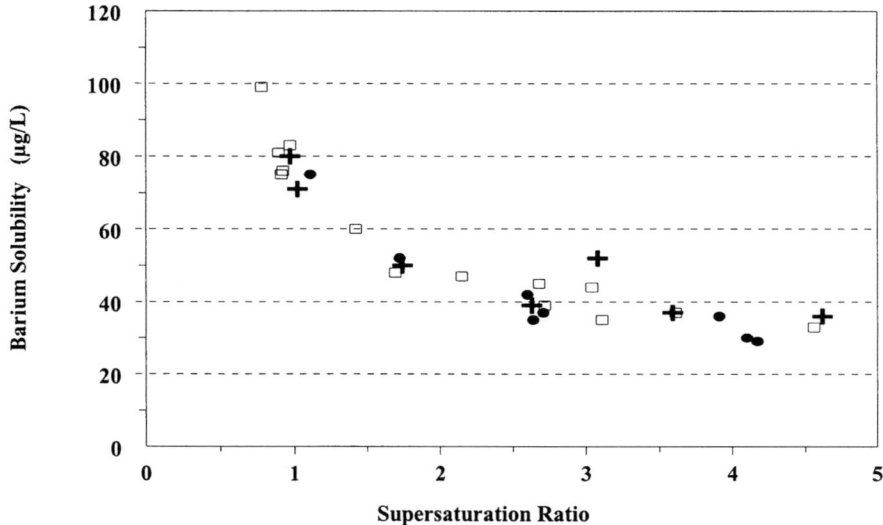

Figure 2.6: Experimental barium solubility in AWS (RO I (□) and RO II (✛)) and synthetic (●) feedwater and concentrate *vs* supersaturation ratio.

2.4.5   Effect of Temperature on BaSO$_4$ Solubility

Data available in literature [22,23] for barium sulphate solubility in pure water at temperatures from 0-90 °C were plotted as a log-log plot of K$_{sp}$ *vs* temperature, Figure 2.7. The linear relation fitted for this data gave the K$_{sp}$ as a function of temperature as follows:

$$K_{sp_T} = \left(\frac{T_{.C}}{25°C}\right)^{0.634} \times K_{sp_{25°C}} \qquad\qquad ...(2.15)$$

where K$_{spT}$ is the thermodynamic solubility product at temperature T (°C) and K$_{sp25°C}$ is the thermodynamic solubility product at 25 °C.

The above relationship was used to correct the Pitzer and Bromley K$_{sp}$ determined in this research for 25 °C to 5 °C. The barium solubility experimentally determined at 25 °C and 5 °C for AWS feedwater and concentrate (sampled on the same day from the first, second and third stage) and synthetic concentrate are compared to Pitzer solubility predictions using the Pitzer K$_{sp}$ at 25 °C and 5 °C (Figure 2.8). No significant difference was found between the solubility predicted by Bromley and Pitzer at 25 °C and at 5 °C. Thus, the line plotted for Pitzer can also be taken to represent Bromley predictions.

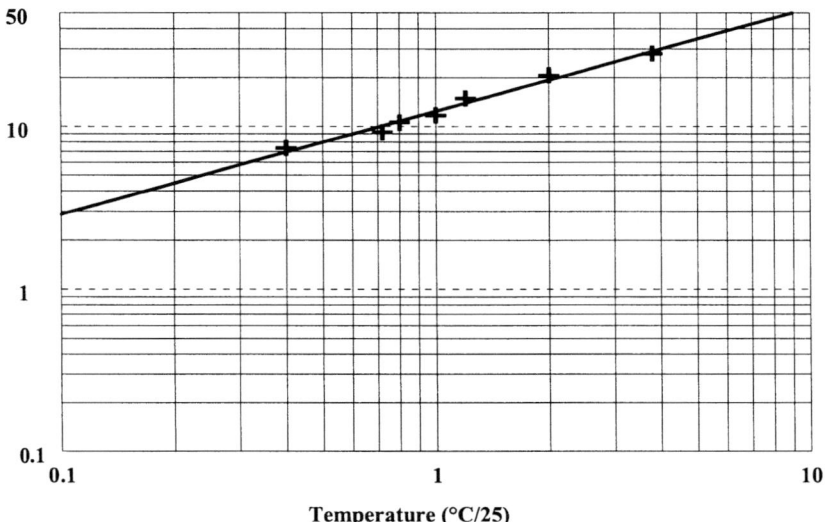

Figure 2.7: Thermodynamic solubility product K_{sp} in pure water [23,24] as a function of temperature. (Slope is 0.634, and intercept ($K_{sp25°C}$) is 1.09×10^{-10} the linear regression coefficient R = 0.990).

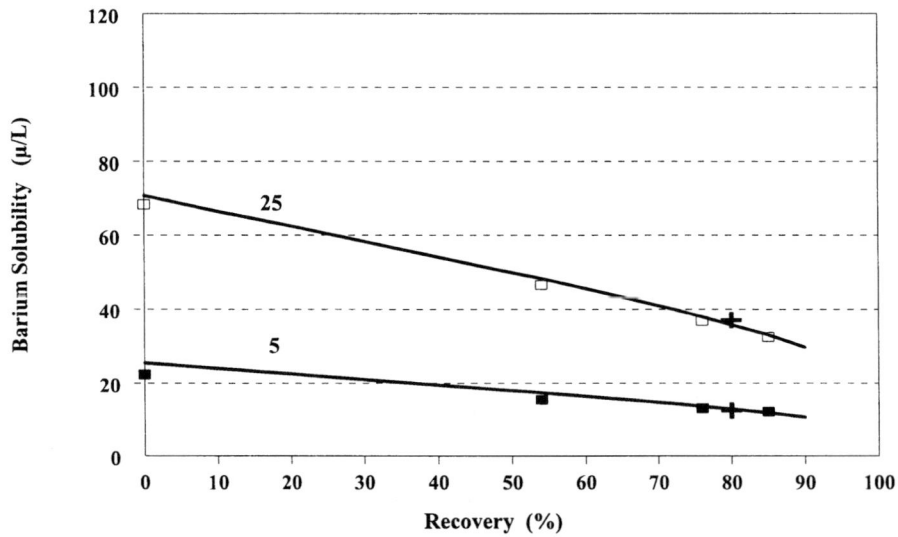

Figure 2.8: Experimentally determined barium solubility of AWS feedwater and concentrate sampled in December 1997 at 25°C (□), and 5°C (■) and synthetic concentrate (✛). Solubility predicted by Pitzer with the experimental K_{sp} for 25°C and K_{sp} adjusted to 5°C.

The significance of a lower temperature in RO operation is in the supersaturation achieved in the concentrate. The pilot plant was operated for one year at 80% recovery with no scaling for temperatures ranging from 0-25 °C and from the results above the solubility is exceeded by a factor of 27 at 5 °C.

2.5 Conclusions

Experimental determination of barium solubility in the RO concentrate confirmed that the RO concentrate are supersaturated, while the feedwater was undersaturated. Solubility prediction was underpredicted by the method in the Du Pont Manual. Accurate prediction was obtained in the two pilot plants examined by both the Bromley Correlation and the Pitzer Model using activities and a K_{sp} calibrated to the RO concentrate. The Bromley method is the easier to apply requiring only the experimental K_{sp} and the ionic strength of the concentrate.

The empirically derived relationship of K_{sp} as a function of temperature was verified allowing accurate prediction at varying temperatures relevant to the drinking water industry for surface water treatment. For more saline solutions the Bromley Correlation overestimates barium sulphate solubility and therefore, would be of limited use for solubility prediction in seawater desalination. Whereas, the Pitzer model can be used for prediction in seawater desalination, but needs the virial coefficients to be optimised and experimental verification.

Organic matter was estimated to potentially complex a negligible amount of barium and found experimentally to have no effect on barium solubility. The major influences on barium solubility were the ionic strength and sulphate concentration (common ion effect). The ratio of sulphate to barium ions, (1000:1) in the AWS concentrate caused an overall decrease in barium solubility, however, the ionic strength effect results in an increase of the barium solubility of 20µg/L. Removal of sulphate ions from the concentrate would increase barium solubility and reduce the overall supersaturation of the concentrate and therefore could play an important role in preventing scaling.

According to the Bromley Correlation and Pitzer Model and the experimentally determined solubilities, the concentrate of the RO system are significantly supersaturated and therefore, thermodynamically, scaling should occur. However, in practice no scaling occurred at 80% recovery and the results of this study confirm that a metastable zone exists in barium sulfate supersaturated solutions for AWS concentrate. This zone is quite wide, *i.e.* at least 27 times the solubility limit at 5 °C. One of the reasons for this wide metastable zone may be the low rate of precipitation of barium sulphate and the effect of organic matter on precipitation *e.g.* through crystal inhibition.

Symbols

A_0 Debye-Hückel constant $(mol/L)^{-0.5}$
a activity (mol/L))
B_1 Bromley ion interaction constant (-)
B_{Ma} second virial Pitzer coefficient for single solute (-)

c_i concentration (mol/L)
c_{eq} barium solubility (mol/L)
C_{Ma} third virial Pitzer coefficient for single solute (-)
CF   concentration factor (-)
F_I modified Debye-Hückel expression(-)
I ionic strength (molar - mol/L solution)/ (Molal - mol/L water)
K_C BaSO$_4$ solubility product (mol/L)^2
K_{sp} BaSO$_4$ thermodynamic solubility product (mol/L)^2
M   Molal (mol/L water)
$M_{Ba}{}^{2+}$ molecular mass of barium (g/mol)
s   solubility (µg/L)
S_r supersaturation ratio (-)
s_C supersaturation (-)
T   temperature (°C)
Y   recovery (-)
z_i ionic charge (-)
Z   constant dependent on ionic charge and concentration (-)

Greek Symbols

$\gamma_\pm$ mean ionic activity coefficient (-)
δ ionic contribution of anion or cation in Bromley ion interaction constant (-)

Subscripts

a   anion
X   anion of interest
c   cation of interest
M   cation

References

1.   J.P. Van der Hoek, P.A.C.Bonné, and E.A.M. Van Soest, Application of hyperfiltration at the Amsterdam Waterworks: Effect of pretreatment on operation and performance, Proceedings of the American Waterworks Association (1995) Membrane Technology Conference, August 13-16 Reno, U.S.A. 277-294.

2.   M.C. Van der Leeden, The role of polyelectrolytes in barium sulfate, Precipitation, Ph.D Thesis, TU Delft, 1991.

3.   J.P. Van der Hoek, P.A.C.Bonné, and E.A.M. Van Soest, Fouling of reverse osmosis membranes: The effect of pretreatment and operating conditions, Proceedings of the American Waterworks Association (1997) Membrane Technology Conference, February 23-26, New Orleans, U.S.A.

4.   Permasep Engineering Manual (PEM). Du Pont de Nemours & Co. 1982.

5.   J.C.Cowan, Water-Formed scale deposits, Gulf Publishing Company, 1992.

6.   J.Buffle, Complexation reactions in aquatic systems, Ellis Horwood Limited 1988.

7.   J.W.Mullin, Crystallization, 3rd Edition. Butterworth-Heinemann, 1993.

8.   O.Söhnel and J.Garside, Precipitation - Basic Principles and Industrial Applications. Butterworth-Heinemann, 1992.

9.   A.S.Myerson, Handbook of Industrial Crystallization. Butterworth-Heinemann, 1993.

10.	J.W.Davis and A.G.Collins, Solubility of barium and strontium sulfates in strong electrolyte solutions. Env. Sci. and Tech. 5, 10, 1971.

11.	I.N.Levine, Physical Chemistry, Third Edition. McGraw-Hill Book Company, 1988.

12.	L.A.Bromley, 1973, Journal of AICh.E 19, 311

13.	K.S.Pitzer, Thermodynamics of electrolytes. I. Theoretical basis and general equations. J. of Phys. Chem., 77 (1973) 268-276.

14.	K.S. Pitzer and G.Mayorga, Thermodynamics of electrolytes. III. Activity and osmotic coefficients for 2-2 electrolytes. J. of Soln. Chem., 3 (1974) 539-546.

15.	R.T.Pabalan and K.S.Pitzer, Thermodynamics of concentrated electrolyte mixtures and the prediction of mineral solubilities to high temperatures for mixtures in the system. Na-K-Mg-Cl-SO$_4$ -OH-H$_2$O Geochemica et Cosmochimica Acta.51 (1987)2429-2443.

16.	F.Van der Ham, Pitzer Model from Laboratory for Process Equipment, Delft University of Technology, Leeghwaterstraat 44, 2628 CA Delft, The Netherlands .

17.	K.S. Pitzer, Activity Coefficients in Electrolyte Solutions, 2nd Edition. CRC Press, 1991.

18.	J.I.Drever, The Geochemistry of Natural Waters. 2nd Edition. Prentice Hall Englewood Cliffs, New Jersey 07632, USA 1988.

19.	M.A.G.T. Van Den Hoop, Electrochemical metal speciation in natural and model polyelectrolyte systems. Ph.D Thesis, TU Wageningen, 1994.

20.	J.I. Kim, G. Buckau, G.H. Li, Duschner, and N. Psarros, Characterization of humic and fulvic acids from gorleben groundwater, Fresenius Journal of Analytical Chemistry, 338 (1990) 245-252.

21.	G.H.Nancollas and N. Purdie, Crystallization of barium sulphate in aqueous solution. Transactions of the Faraday Society, 59 (1963) 735-740.

22.	C.C.Templeton, Solubility of barium sulphate in sodium chloride from 25° to 95°C. Journal of Chemical and Engineering Data, 5 (1960) 514-516.

23	W.R.Linke, and A. Seidell, Solubilities - inorganic and metal-organic compounds. Vol II American Chemical Society, Washington DC, 1965.

3

Stable Barium Sulphate Supersaturation in Reverse Osmosis

Chapter 3 is based on: "Stable Barium Sulphate Supersaturation in Reverse Osmosis" by S.F.E. Boerlage, I. Bremere, M.D. Kennedy, G.J. Witkamp, J.P. Van der Hoek and J.C. Schippers. Published in *Journal of Membrane Science*, (2000) Vol. 179, pp. 53-68.

Contents

Abstract

A RO pilot plant operated without antiscalant addition at 85% recovery with no scaling, although the concentrates were significantly supersaturated with barium sulphate. Stable supersaturation may be due to slow precipitation kinetics which may be retarded or enhanced by organic matter present in RO concentrate. Barium sulphate precipitation kinetics; crystal nucleation, measured as induction time, and growth were investigated in batch experiments in RO concentrate and in synthetic concentrate containing (i) no organic matter and (ii) commercial humic acid. Supersaturation appeared to control induction time. Induction time decreased more than 36 times with a recovery increase from 80% to 90%, corresponding to a supersaturation of 3.1 and 4.9, respectively. Organic matter in 90% RO concentrate did not prolong induction time (5.5 hour). Whereas, commercial humic acid extended induction time in 90% synthetic concentrate to >200 hours. This was most likely due to growth inhibition as growth rates determined by seeded growth in synthetic concentrate containing commercial humic acid were reduced by a factor of 6. In comparison, growth rates were retarded only 2.5 times by organic matter in RO concentrate. However, growth rates measured for 80 and 90% RO concentrate were significant and not likely to limit barium sulphate scaling. Results indicate that the nucleation rate expressed as induction time is governing the occurrence of scaling.

3.1   Introduction

Reverse osmosis (RO) has become a cost effective and viable technology in drinking water production for surface, brackish and seawater sources. However, the precipitation of sparingly soluble salts on the membrane, referred to as scaling, is widely recognised as a serious problem in reverse osmosis and nanofiltration applications. Scaling may result in a decline in membrane water production and potentially membrane failure. Barium sulphate (barite) is particularly troublesome in reverse osmosis due to its low solubility (1×10^{-5} mol/L in pure water) [1]. Furthermore, if it is not detected at an early stage and has aged into a hard deposit, barite scale is resistant to removal by conventional cleaning chemicals.

The method in the Du Pont Manual is the most widely applied method in RO systems to predict barium sulphate scaling. This method predicts scaling if the solubility of the salt is exceeded at the design recovery [2]. If this is the case, antiscalant addition or lowering the recovery is generally recommended to prevent scaling. However, lowering the recovery is not a financially desirable option and antiscalant addition involves extra operational costs. In addition, the discharge of membrane concentrates without antiscalants is more environmentally favourable.

Accurate scaling prediction is essential in RO in order to maximise recovery and to determine if antiscalants are really necessary. Problems associated with inaccurate scale prediction include: underestimation of a feedwater's scaling potential resulting in scaled membranes or its overestimation, resulting in antiscalant overdosing.

This work focuses on the Amsterdam Water Supply (AWS) RO pilot plant where RO is under investigation as an integrated part of two treatment systems. In the treatment system relevant to this research, the pretreated River Rhine water undergoes further treatment by ozonation - biologically activated carbon filtration - slow sand filtration prior to reverse osmosis [3].

The RO pilot plant was initially operated at 90% recovery with sulfuric acid addition to prevent $CaCO_3$ scaling [3]. According to the Du Pont method the feedwater (before acid addition) was already 1.5 times saturated with barium sulphate. Therefore, an antiscalant was added, nevertheless, barium sulphate scaling occurred. Changing the antiscalant solved the scaling problem, however, biofouling occurred due to the antiscalants biodegradability [4].

Consequently, the recovery was lowered to 80%, hydrochloric acid replaced sulfuric acid addition (increased supersaturation by addition of sulphate ions) and no antiscalant was added. Surprisingly, no scaling occurred even though the concentrates were confirmed in Chapter 2 to be significantly supersaturated with barium sulphate; 9 times the solubility at 25 °C [5]. The recovery was therefore increased to 85% and *safe* operation (no scaling) was maintained for an entire year of operation [6].

In practice, the existing Du Pont method for barium sulphate scaling prediction is deficient in many respects. In AWS seasonal temperature variations can range from 24 °C in summer to close to 0 °C in winter. However, as the Du Pont method is limited to 25 °C the real supersaturation in the RO plant can not be accurately quantified, as barite solubility varies dramatically with temperature [7]. Therefore, a method was developed in Chapter 2 which accurately quantifies supersaturation in RO concentrate in the temperature range 5-25 °C. Employing this new method, supersaturation was found to be almost three times greater at 5 °C in the AWS 85% concentrate in comparison to 25 °C. Nevertheless, solubility based methods cannot accurately predict when scaling will occur nor account for the stable supersaturation phenomenon.

The stable supersaturation observed at the RO pilot plant may be explained by *metastability* of the concentrate. In literature, a metastable solution refers to a saturated solution where precipitation is thermodynamically possible but spontaneous precipitation is improbable due to slow kinetics [8,9]. Precipitation kinetics comprises two steps; firstly *nucleation*; "birth" of a new crystal followed by crystal *growth*. Either of these two steps may dictate when precipitation or scaling occurs. Supersaturation is the driving force for both processes, the higher it is the more favourable precipitation becomes [8]. Similarly, an increase in temperature will increase both nucleation and growth, in addition to having an effect on supersaturation through solubility.

Alternatively, inhibition of precipitation kinetics by organic matter present in the surface water concentrate may occur resulting in stable supersaturation. Organic compounds of both natural and anthropogenic origin in surface water may have a profound effect on scaling. Humic acids, which can comprise up to 50% of the dissolved organic carbon in river water [10, 11], are known to adsorb on the surface of crystals which could lead to growth inhibition in the same way as antiscalants [12].

This study investigates the cause of stable supersaturation in the RO concentrate. The barium sulphate nucleation rate, measured as induction time, and the growth rate as a function of supersaturation in the RO concentrate are examined. In particular, the effect of natural organic matter and commercial humic acid, chosen as a model organic compound, in inhibiting nucleation/and or growth in concentrate will be investigated.

3.2   Background

3.2.1   Precipitation Kinetics

Supersaturation can be quantified by the supersaturation ratio S_r [13]. For $BaSO_4$ this is calculated as the product of the concentration of the barium and sulphate ions and their activity coefficients, which takes into account possible ion associations in solution, divided by the thermodynamic solubility product K_{sp};

$$S_r = \sqrt{\frac{\gamma_+ [Ba^{2+}] \gamma_- [SO_4^{2-}]}{K_{sp_T}}} \qquad \qquad ...(3.1)$$

3.2.1.1   Nucleation

Nucleation refers to the formation of a new phase, the crystal solid. Once supersaturation occurs, lattice ions start to associate and form nuclei or clusters. Below a critical size the probability that the cluster survives is limited, therefore, these clusters typically redissolve. When the cluster grows beyond the critical size, it becomes more stable and can subsequently grow.

Various mechanisms of nucleation exist. Primary nucleation occurs in the absence of the crystalline solid being formed which includes; *homogeneous* (spontaneous), dominant at high supersaturation and *heterogeneous,* induced by foreign particles or dissolved system impurities effective at lower supersaturation. Secondary nucleation involves the presence of the crystalline solid being formed itself which produces a catalysing effect on nucleation [9,13].

Homogeneous nucleation forms the basis of several nucleation theories. Two important parameters characterize the nucleation process: (i) the free energy change, and (ii) the rate of nuclei formation. The nucleation rate, J_n (nuclei/cm^{3}s), in classical theory is given by an Arrhenius type of expression [8,14]:

$$J_n = A \exp\left[\frac{-\Delta G_{cr}}{kT}\right] \qquad \qquad ...(3.2)$$

where A is the pre-exponential factor related to the efficiency of collisions of ions and molecules, which has a value of *circa.* 10^{23} to 10^{33}, k is the Boltzmann constant, T is the absolute temperature, and ΔG_{cr} is the free energy change for the critical cluster size to form. The nucleation rate can be expressed in molar quantities [14, 15]:

$$J_n = A \exp\left[-\frac{F V_m^2 \sigma^3 f(\theta) N_a}{(RT)^3 (\ln S_r^2)^2}\right] \qquad \qquad ...(3.3)$$

where F describes the crystal geometry (relating area and volume shape factors *e.g.* for a spherical nucleus $16\pi/3$), V_m is the molar volume of solid, σ the interfacial tension, $f(\theta)$ contact angle factor, N_a is Avogadro's number, and R is the universal gas constant.

The nucleation rate, J_n, is critically dependent on supersaturation (refer Equation 3.3). As supersaturation increases the critical cluster size decreases and the probability that clusters will survive to form crystals increases. Thus, with increasing supersaturation, nucleation is more energetically favourable and eventually, nucleation occurs spontaneously. Whereas, below a certain supersaturation the nucleation rate is virtually zero or so slow as not to be observed within the measured time frame, this is referred to as the *metastable* region [8,16]. Similarly, from Equation 3.3 a temperature increase will increase nucleation.

Classic nucleation theory can be extended for heterogeneous nucleation. In homogeneous nucleation the contact angle factor $f(\theta)$ is 1 whereas, in heterogeneous nucleation impurities can act as a surface or nuclei for nucleation and $f(\theta)$ typically decreases [15]. This reduction is due to increased wettability of the new phase on a foreign nucleus which promotes nucleation and can reduce the width of the metastable region [8]. In addition, impurities and other inorganic ions may increase or decrease surface tension (σ) and consequently increase or decrease the nucleation rate in heterogeneous nucleation [1]. Thus, heterogenous nucleation can be inhibited by adsorption of impurities. However, inhibition of nucleation is extremely rare and at higher supersaturation homogenous nucleation dominates and this cannot be prevented.

In practice the concept of *induction time* is frequently used to describe nucleation. The induction period, t_{ind}, is defined as the time elapsed between the creation of supersaturation and the first appearance of a new phase, ideally nuclei with the critical cluster size dimensions [1,17]. The induction time may be measured by following the change in concentration of one of the crystal ions over time (refer nucleation section of Figure 3.1). However, as the induction period is determined experimentally, its accuracy in approximating nucleation *depends* on the analytical technique used. Consequently, a part of the induction time may also include growth to a detectable size [13]. For the purposes of this research it is assumed that the nucleation time is much greater than the time required for growth of crystal nuclei to a detectable size. Thus, induction time can be assumed to be inversely proportional to the rate of nucleation, $t_{ind} \propto J_n^{-1}$ and, Equation 3.3 may be written as [9,13]:

$$\log t_{ind} = \frac{B}{(\log S_r^2)^2} - E \qquad\qquad ...(3.4)$$

where a linear relation can be found for $\log t_{ind}$ and $(\log S_r^2)^{-2}$ where E is a constant and B the slope of the line is:

$$B = \frac{F V_m^2 \sigma^3 f(\theta) N_a}{(RT)^3} \qquad\qquad ...(3.5)$$

Typically, the dependence observed for a wide range of supersaturation can be separated into two linear regions with different slopes [15]. This is due to a change in the nucleation mechanism, one region corresponds to homogeneous, and the other heterogeneous. These two regions can be connected by a continuous curve [13].

An empirical relationship was suggested by Gunn [18] for the relationship of induction time with temperature, at a constant supersaturation, which is similar to the Arrhenius equation for the temperature dependence of the rate constant:

$$\log \frac{1}{t_{ind}} = A_1 - \frac{E_a}{2.303\ RT} \qquad\qquad ...(3.6)$$

where E_a is the activation energy for nucleation and A_1 is a constant. Gunn [18] found a linear relationship for a $BaSO_4$ supersaturation ratio of 20 and 29 when the temperature varied between 17-50°C. This relationship has been tested for other salts and a linear relationship was found for example with calcium sulphate dihydrate [19, 20] in the latter case from 25-90°C.

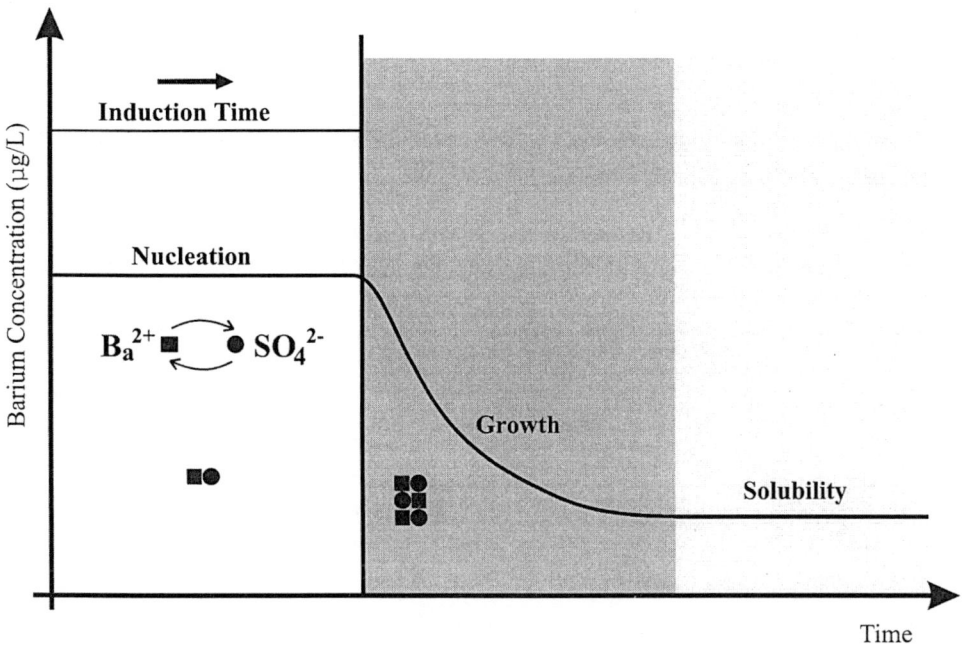

Figure 3.1: Barium concentration over time showing (i) a stable region where nucleation is estimated by the experimental determination of induction time followed by (ii) barium desuperaturation due to growth of nuclei, from spontaneous nucleation or induced by seed crystal addition, to final equilibrium saturation.

3.2.1.2   Crystal Growth

After nucleation the crystal nuclei grow by addition of crystal ions from the supersaturated solution continuing until solution equilibrium (solubility) is reached. To avoid difficulties in the reproducibility of crystal growth associated with spontaneous precipitation the seeded growth technique was introduced [21]. This involves the inoculation of the supersaturated solution with a suspension of small seed crystals of the salt concerned. The growth process is then followed by registering the change in the bulk concentration of one of the crystal ions in solution as a function of time.

The general growth process is shown in Figure 3.1 for barium and is commonly referred to as a desupersaturation curve. Initially, if nucleation has occurred or seed crystals are added a very rapid concentration decrease (corresponding to rapid growth) is recorded due to the large seed crystal surface available coupled to the high initial supersaturation. This is followed by a region of slower change in concentration to reach equilibrium.

The overall growth rate describes the rate of change of the seed crystals volume dV/dt at time t and can be calculated in seeded growth experiments from the desupersaturation curve [22]:

$$\frac{dV}{dt} = \frac{V_s \, M_{BaSO_4}}{\rho} \frac{dc}{dt} \qquad \qquad ...(3.7)$$

where V_s is the volume of solution, dc/dt is the decrease in the crystal ion over time, M_{BaSO4} and ρ are the molar mass and density of the salt, respectively.

The mean linear growth rate G_{lin}, defined as the growth rate of a crystal face in the direction normal to the face, is often employed in characterising the growth of sparingly soluble salts such as $BaSO_4$ [1, 23]:

$$G_{lin} = - \frac{\dfrac{dV}{dt}}{A_{(t)}} \qquad \qquad ...(3.8)$$

where $A_{(t)}$ is the total surface area of the seed crystals at time t, calculated by Equation 3.9:

$$A_{(t)} = A_{(0)} \left[\frac{m_{(t)}}{m_{(0)}} \right]^{\frac{2}{3}} \qquad \qquad ...(3.9)$$

For sparingly soluble salts, the outgrowth factor $m_{(t)}/m_{(0)}$ (crystal mass at time t /crystal seed mass at time zero) is small and can be neglected thus, $A_{(t)}$ remains essentially unchanged [1, 22]. The dependence of the linear growth rate on supersaturation can be expressed by the empirical rate law [1]:

$$G_{lin} = K_g \, (S_r - 1)^p \qquad \qquad ...(3.10)$$

where p is the order of the reaction, and K_g is the growth rate constant. Extensive data published on growth rate studies of $BaSO_4$ indicate that growth is second order with respect to supersaturation and the growth process is surface integration controlled [1, 21, 24-26]. The relationship of G_{lin} is often investigated by plotting G_{lin} as a function of supersaturation expressed as growth affinity β. Where β is defined as a function of the difference in the chemical potential ($\Delta\mu$) of a growth unit in the supersaturated solution and in its crystalline form divided by RT [1, 22]. If the linear growth rate depends *only* on supersaturation then all curves should coincide in a plot of G_{lin} vs β at a constant temperature [23].

$$\beta = - \frac{\Delta\mu}{RT} = 2 \ln S_r \qquad \qquad ...(3.11)$$

Although the growth rate constant does not depend on supersaturation it is temperature dependent and usually fits an Arrhenius equation, Therefore, the growth rate as a function of temperature and supersaturation is [8]:

$$G_{lin} = A_2 \exp^{-\frac{E_a}{RT}} (S_r - 1)^p \qquad ...(3.12)$$

where A_2 is a constant, and E_a is the activation energy.

3.3 Materials and Methods

3.3.1 Materials

Analytical reagent grade chemicals and deionised water were used in the experiments. Barium standards, the certified barium standard to check the accuracy of calibration and labware were prepared and employed as described in Chapter 2. Commercial barite crystals (Baker Analyzed) were used as seed crystals. The seed crystals (10mg) were placed in 1mL deionised water for 24 hours prior to use. Their mean specific surface area was determined from BET gas adsorption measurements (Quantasorb, N_2/He mixtures) to be $6.32\pm0.04m^2/g$ (triplicate measurement).

3.3.2 RO concentrate

The barium and sulphate concentration in the RO feedwater is 40-90µg/L and 40-80mg/L, respectively. Samples were taken from the RO pilot plant (RO I line) at various recoveries between 80-90%. Sampling was limited to 90% recovery to prevent any scaling occurring in the pilot plant. The pilot plant consists of seven pressure vessels in a three stage configuration (4-2-1) and produces $9.1m^3/h$ at 90% recovery. At the time of sampling hydrochloric acid was added (to give an LSI in the concentrate of 0) to prevent $CaCO_3$ scaling and no $BaSO_4$ antiscalant was added.

3.3.3 Blank Synthetic Concentrate

Synthetic water was prepared as described in Chapter 2 to reach the same chemical composition and pH as AWS concentrate at 80 and 90% recovery. Blank synthetic concentrate contained a negligible amount of organic matter ($\leq0.1mg/L$ DOC).

3.3.4 Synthetic Concentrate with Humic Acid Addition

In the experiments with addition of commercial humic acid (Acros), it was assumed that the dissolved organic carbon present in the RO feedwater (1 mg/L DOC) is principally humic acid.

3.3.4.1 Humic Acid Preparation

The humic acid sodium salt (10g) was dissolved in 1L of deionised water at ambient temperature. To remove the sodium a multiple washing procedure, using deionised water, was performed in which the humic acid was reprecipitated with HCl (6M) prior to centrifuging. The solution was centrifuged (Homef LC-30) in small portions for 10 minutes, the supernatant was discarded and the sediment was redissolved in deionised water. The centrifugation-redissolution procedure was repeated two more times. The sediment was then redissolved and centrifuged. The supernatant was decanted and subsequently filtered through an acid washed glassfiber filter (Whatman GF/C pore size 1.2μm). The filtrate collected, referred to hereafter as the humic acid standard solution, was yellowish in colour, clear and with no particulate material visible. The DOC in the obtained humic acid standard solution was measured (TOC Analyzer Model 700, O-I Corporation) to estimate the required volume to be added to the respective synthetic concentrate. Table 3.1 summarises the amount of humic acid as DOC added in the growth and induction time experiments in the 80 and 90% synthetic concentrate.

Table 3.1:   Addition of humic acid standard expressed as DOC (mg/l) to synthetic concentrate at 80 and 90% recovery in induction time and growth experiments.

Experiment	Humic Acid Addition as DOC (mg/L) to Synthetic Concentrate	
	80 % Recovery	90 % Recovery
Induction Time	10	20
Growth Rate	5	10
	10	20

3.3.5   Experimental Procedure

Batch test experiments, with and without seed crystal addition, were used to examine the barium sulphate growth rates and induction times in the various concentrate, respectively.

Batch reactors (1L closed plastic bottles) containing RO and synthetic concentrate, with and without humic acid addition, were pre-equilibrated to 25°C, and shaken on a Rotator-Model G2 at 200rpm for the experiment duration at 25°C. In seeded growth experiments, growth was initiated by the addition of the seed crystal dose (in suspension) which corresponds to time zero. For induction time no seed crystals were added and time zero corresponded to the time of agitation after temperature pre-equilibration. Experiments were carried out in duplicate. Periodically 10mL samples in duplicate were taken for barium analysis from the batch reactors using a plastic syringe (Becton Dickinson) with a 0.2μm filter (Sartorius) rinsed with 0.2M HNO_3, prior to use. The first 2mL of sample filtered was discarded and the following 8mL of filtrate collected, 1 drop of concentrated (14 M) HNO_3 was added to each sample. The bottle was manually shaken between the withdrawal of each 10 mL sample to keep the seeds in suspension.

To follow the growth and nucleation processes (induction time) the concentration of barium in the samples over time was determined by Inductively Coupled Plasma Spectrophotometry (ICP) (Spectroflame 21/016) according to Chapter 2.

3.3.6   Data Analysis

3.3.6.1   Growth Rate Determination

Samples were collected every 10 minutes for the first three hours and thereafter every hour, in the seeded growth experiments. The mean linear growth rate was determined according to Equation 3.8 from the desupersaturation curve generated from the results of barium over time. Due to the reduction in volume with sampling, the batch test experiments can not be considered as a constant volume reactor. However, it is assumed that the bulk supersaturated solution and seed crystals constitute a homogenous suspension. Therefore, when a sample is taken, crystals are withdrawn together with a proportional amount of solution. Thus, a correction for a reduction in volume and seed crystal area would cancel in Equation 3.8 and no adjustment is required.

The slope of the best fit straight line for the first hour of the desupersaturation curve (dc/dt) was considered as representative of the high initial supersaturation and used to calculate the *initial linear growth rate* and the growth rate constant (Equation 3.10). The outgrowth for the initial linear growth rate was calculated to be less than 3%, which results in a surface area change of less than 1.3% (Equation 3.9), and can therefore be neglected. In experiments where the linear growth rate in a concentrate was examined over time, the outgrowth and surface area was estimated to increase by maximum 10 and 6%, respectively. As most of the data presented in these graphs corresponds to a surface area change of <5% it is assumed to be within experimental error.

3.3.6.2   Induction Time Determination

Variations in the barium concentration were monitored over time in unseeded growth experiments. The induction time was measured experimentally as the period of stability before an observable change was detected by ICP.

3.3.6.3   Calculation of Supersaturation Ratio

The supersaturation ratio of the RO and synthetic concentrate was calculated using Equation 3.1. The thermodynamic K_{sp} determined previously for the RO concentrate of 1.34×10^{-14} was employed for 25 °C. Activity coefficients in this research were calculated by the Blue Sea programme [27] which uses Pitzer equations as described in Chapter 2, however, as demonstrated in Chapter 2 the Bromley correlation could equally have been employed.

3.4   Results and Discussion

3.4.1   Nucleation

3.4.1.1   Effect of Organic Matter

To establish if stable supersaturation was due to slow nucleation kinetics, in particular due to the effect of organic matter present in the concentrate, the stability of RO and synthetic concentrate over time was examined. Results of the RO concentrate and synthetic concentrate containing; (i) no organic matter (blank) and (ii) humic acid (equivalent to 2mg/L in the feedwater) at 80% and 90% recovery are presented in Figure 3.2 and 3.3, respectively. At 80% recovery all the concentrates appeared stable over the 200 hours of the experiment with no change in barium concentration observed. Thus, the induction time was greater than 200 hours.

In contrast, at 90% recovery (refer Figure 3.3) the barium concentration in RO (10mg/L DOC) and blank synthetic concentrate ($\leq$0.1mg/L DOC) decreased over time at a similar rate and the induction time for both was 5-6 hours. Therefore, the organic matter present in the RO feedwater did not appear to markedly inhibit nucleation.

However, the 90% synthetic concentrate containing humic acid (20mg/L DOC) remained stable for the entire experiment duration. This may be attributable to an increase in the interfacial tension (σ) in heterogeneous nucleation. He et al [28] found that the addition of a selected inhibitor increased the interfacial tension between the crystal and aqueous solution by a factor of 1.8 which prolonged the induction time. The humic acid tested here may have retarded heterogeneous nucleation in a similar way. Although, it is more likely that the humic acid adsorbed onto the freshly formed crystal nuclei and inhibited growth to a detectable size.

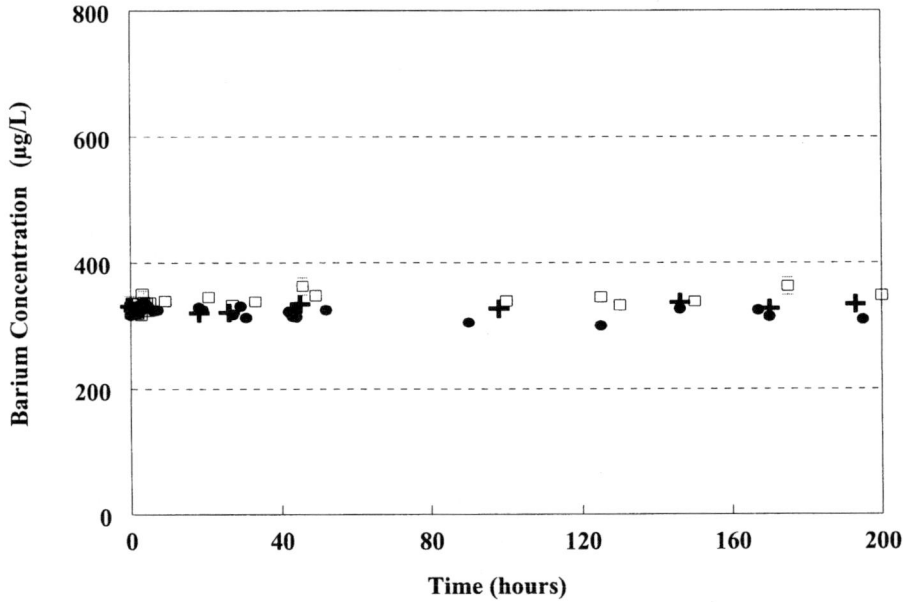

Figure 3.2:   Concentration of barium over time for 80% recovery; RO concentrate (5mg/L DOC) □ and synthetic concentrate containing (i) no organic matter ($\leq$0.1mg/L DOC) ● and (ii) 10mg/L humic acid DOC ✚.

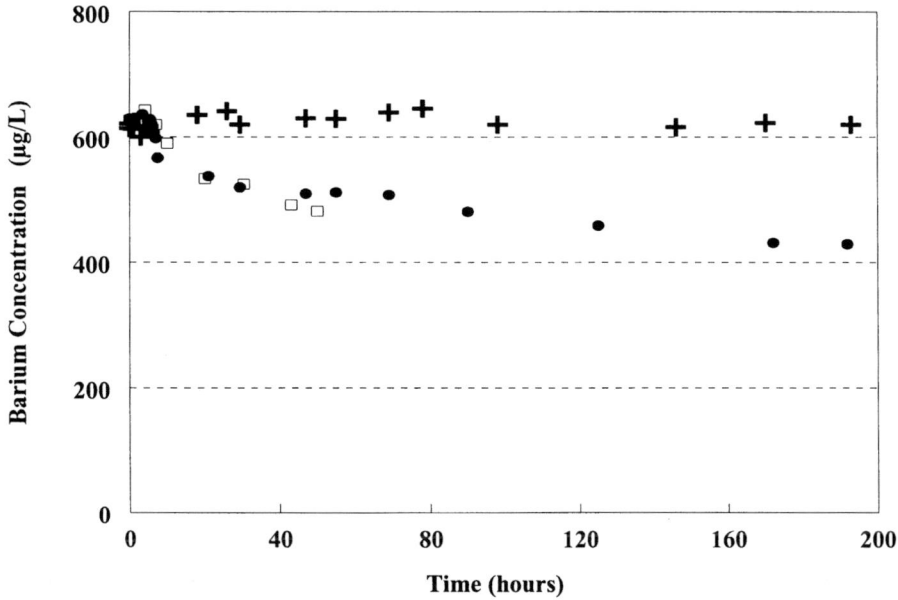

Figure 3.3: Concentration of barium over time for 90% recovery; RO concentrate (10mg/L DOC) □ and synthetic concentrate containing (i) no organic matter (≤0.1mg/L DOC) ● and (ii) 20mg/L humic acid DOC ✚.

Results from this section indicate that nucleation is not inhibited by organic matter present in the RO concentrate. The difference in effect of RO organic matter and commercial humic acid on induction time may be due to the lower DOC concentration and/or that not all the DOC in the RO concentrate is present as humic acid. Therefore, simply not enough humic acid is present in the RO concentrate to prevent nucleation. Alternatively, the type and behaviour of RO organic matter may differ to that of the commercial humic acid. Conflict in literature over the use of commercial humic acid is well known. It is argued that commercial humic acid is not representative of aquatic humic acid and is of limited use [29]. Whereas, it has also been reported that despite some differences between commercial humic acid and isolates from the environment the chemical reactivity was found to be similar [30].

3.4.1.2   Effect of Supersaturation

In Figure 3.4 the stability of RO concentrate in the recovery range of 86-90% is presented. All the concentrates showed a barium concentration decrease after a period of stability where the higher the recovery the shorter the stable region observed. As recovery increases supersaturation increases and this is the key factor determining the stability of the RO concentrate. Induction time determined for RO concentrate at various recoveries with the corresponding supersaturation (expressed as supersaturation ratio) and organic matter content (DOC) for each recovery is given in Table 3.2. The induction time is observed to decrease by a factor of 36 when the recovery is increased from 80 (S_r=3.1) to 90% (S_r=4.9), recovery (refer Table 3.2).

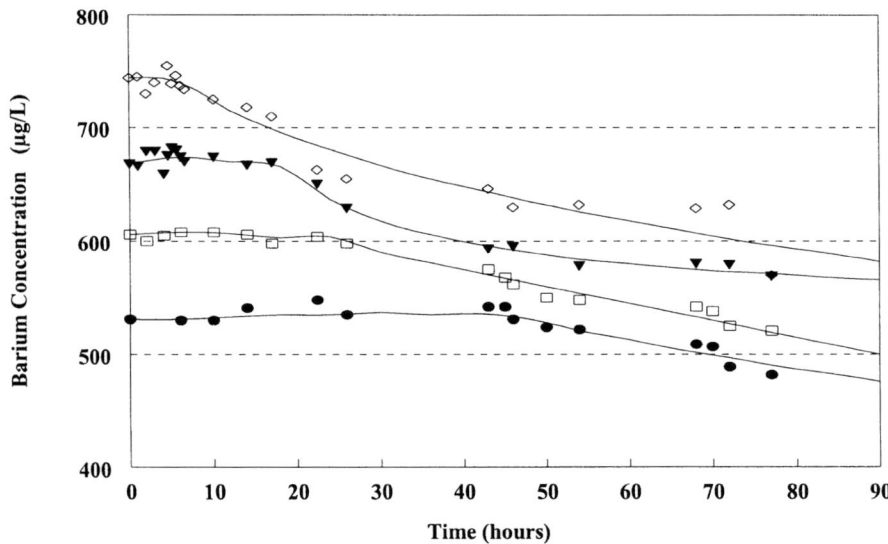

Figure 3.4: Barium concentration over time for RO concentrate as a function of recovery: 90% (◇), 89% (▼), 88% (□), 86% (●).

Table 3.2: Induction time at 25°C for RO concentrate at 80-90% recovery with organic matter concentration expressed as dissolved organic carbon and supersaturation ratio.

Recovery (%)	Estimated DOC from RO feedwater 1 mg/L (mg/L)	Supersaturation Ratio, S_r (-)	Induction Time (hr)
90	10.0	4.94	5.5
	10.0	4.56	8.0
89	9.1	4.64	17
88	8.3	4.25	23
86	7.1	3.76	46
80	5.0	3.11	>200

The dependence of induction time on supersaturation for the RO concentrate was examined in the $\log t_{ind}$ *vs* $(\log S_r^2)^{-2}$ plot (refer Figure 3.5). The induction time of $BaSO_4$ has been extensively studied in literature in synthetic solutions of various supersaturation at 25°C [1,17,31]. Nucleation was considered as predominantly heterogeneous due to the presence of impurities [1,17,31]. Data from these studies is included for comparison in Figure 3.5 for a supersaturation ratio below 30 as He *et al* [17] reported a change in the nucleation mechanism from heterogenous to homogenous at this supersaturation.

It is evident that the RO induction time (S_r 3.7-5.0) follow the trend found in literature, although most reported data is for a higher range of supersaturation (S_r 4.4-30.0). A function was fitted to the induction time determined for the RO concentrate and extrapolated for higher supersaturation. The fitted function for higher S_r lies between the data plotted from literature for pure water [1,31] and 1M NaCl, [17] as expected.

Nucleation in the RO concentrate appears to be limited by long induction times. The function fitted in Figure 3.5 could then be used to calculate induction time for RO concentrate at higher supersaturation in order to estimate safe operating conditions to prevent scaling.

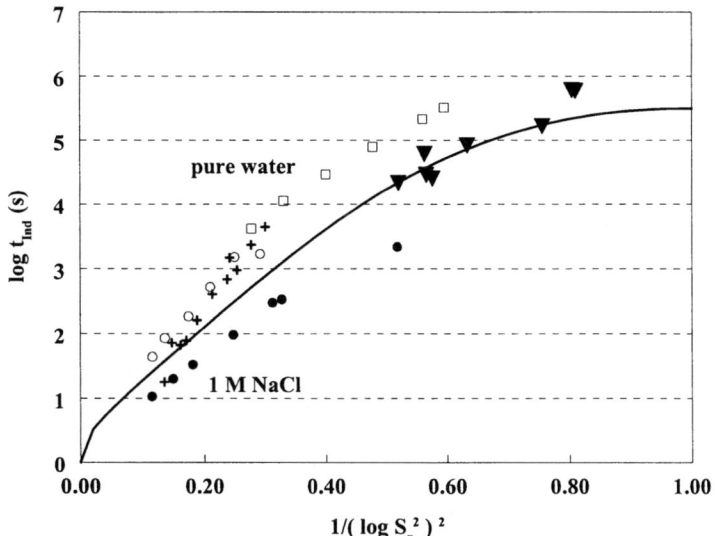

Figure 3.5: Effect of supersaturation on induction time at 25°C for RO Concentrate (▼), compared with published values in pure water (□) [1], ✚ [31]; and in sodium chloride solutions [17]; dilute <0.009M (○) and 1M sodium chloride (●). Best fit function of results for RO concentrate (-).

3.4.1.3   Effect of Impurities, Ionic Strength and BaSO₄ Stoichiometry

In the case of RO concentrate, nucleation is expected to be heterogenous due to the low supersaturation range and the presence of organic matter and colloidal particles such as alumino silicates in river water. The synergistic effect of these system "impurities" on the measured induction time of the RO concentrate can be interpreted generally according to classical nucleation theory. The parameters at a constant supersaturation and temperature that affect the nucleation rate and hence the induction period are the interfacial tension (σ) and the contact angle factor $f(\theta)$ for heterogenous nucleation (refer Equation 3.3).

In Figure 3.5, the induction time of RO concentrate were found to be shorter than those reported by van der Leeden [1] for a similar supersaturation in pure water, which is most likely due in part to the catalysing effect of particulate matter reducing $f(\theta)$. Colloidal particles such as Al_2O_3 were shown to lower nucleation energies and increase nucleation rates [32]. This effect can be amplified at higher recoveries due to the higher concentration of particles which promote nucleation and at higher supersaturation a large variety of foreign particles can act as effective nucleants. Whereas, at lower supersaturation closer matching of the particle to the crystal ion is required for the release of supersaturation [33].

Assuming that $f(\theta)$ is 0.4, based on Sohnel and Mullin [15], the interfacial tension in the RO concentrate was estimated from the slope (Equation 3.5) of the fitted curve in Figure 3.5 for the RO concentrate to be $48mJ/m^2$. The low value supports heterogenous nucleation, and is close to the estimate of $40mJ/m^2$ reported by van der Leeden [34] at $25°C$ for pure water but is lower than that of $84mJ/m^2$ also reported for pure water by Symeopoulos et al [31].

Increasing the ionic strength of a solution may also reduce the interfacial tension and result in lower induction times. He et al [17] reported a decrease in induction time (refer Table 3.3) when the ionic strength was increased from 0.0002 to 1M for a constant supersaturation ($S_r = 10$). The induction time measured by He et al is a factor 2.5 times shorter for a ten fold increase in ionic strength from 0.01 to 0.1M which corresponds to the ionic strength of the RO feedwater and 90% concentrate, respectively. He et al attributed the catalytic effect of ionic strength to increasing solubility with increasing ionic strength, which is correlated with decreased interfacial tension [13,17]. However, although the ionic strength increases ten times in the RO feedwater to 0.10M in the 90% RO concentrate, solubility was found to decrease with ionic strength (refer Chapter 2 section 2.4.3). Nevertheless, ionic strength may contribute to the lower value of the interfacial tension in the RO concentrate and cause a decrease in induction time.

Table 3.3:   Effect of Ionic Strength increase on Induction time at $25°C$ for synthetic solutions at a constant supersaturation ratio of 10 reported by He et al [17].

Ionic Strength (M)	Induction Time (min)
0.0002	25.2
0.01	15.2
0.1	6.0
0.5	3.2
1	1.7

Another difference in the RO concentrate from the literature data presented in Figure 3.5 is the stoichiometry of the barium and sulphate ions. In most surface waters the sulphate ion is present far in excess of the barium ion and in the case of the RO feedwater used in this study it is 1000:1 (refer Chapter 2). However, in pure water solutions, Symeopoulos et al [31] found no effect for a SO_4^{2-} to Ba^{2+} 100:0.01 ratio and He et al [28] found only a slight dependence for the 500:0.005 ratio. Therefore, this is not expected to have an effect on induction time.

3.4.1.4   Effect of Temperature

A temperature or supersaturation increase will increase nucleation and hence the potential for scaling in the RO system. However, a temperature increase will also lower supersaturation due to an increase in solubility [5,7]. Therefore, the stability of membrane concentrate can vary dramatically with seasonal temperature changes. A method exists for the prediction of barium sulphate solubility as a function of temperature but a means to correct induction time for temperature effects is required.

The most systematic study of the effect of temperature on $BaSO_4$ induction time was carried out by He $et\ al$ [17]. In this study He $et\ al$ demonstrated linearity for a S_r of 10 using the empirical relationship given in Equation 3.6 for the temperature range of 25-90°C in 1M NaCl solutions. To examine the effect of temperature on $BaSO_4$ induction time for the supersaturation range relevant to the RO concentrate, data reported by He $et\ al$ [17] were linearised in the plot of log (t_{ind}^{-1}) vs 1000/T.

The results for five chosen supersaturation ratios; 5, 10, 15, 20, 30 and for the temperature range 25-90°C available in He $et\ al$'s study are presented in Table 3.4. The slope and intercept of the line obtained for each supersaturation was employed to predict the induction time at 25°C and 5°C (range verified for the barium solubility prediction method) for the 1M NaCl solutions, assuming that linearity holds at this lower temperature.

Table 3.4:   Comparison of predicted induction time for $BaSO_4$ in 1M NaCl solutions at 5 and 25°C using the slope and intercept obtained from plotting data reported by He et al [17] as the log reciprocal induction time vs temperature relationship for a given supersaturation.

Supersaturation Ratio S_r	Slope	Intercept	Predicted Induction Time (min)	
			5°C	25°C
5	-2.31	4.41	131.1	36.3
10	-2.16	5.23	5.8	1.7
15	-1.66	4.05	1.4	0.6
20	-1.34	3.28	0.6	0.3
30	-1.08	2.65	0.3	0.2

It is evident from Table 3.4 that the effect of temperature on induction time diminishes with increasing supersaturation as seen by the decrease in the slope. For example, at a S_r of 5 the slope is -2.31 and at 30, -1.08. Thus, the predicted induction time at a supersaturation ratio of 5 for a temperature change of 20°C (from 25 to 5°C) extends the induction time by a factor of 3.6. Whereas, for the same temperature change at the higher supersaturation ratio of 30 the induction time is only increased by a factor of 1.5. This is in agreement with Equation 3.3 whereby the effect of supersaturation and temperature on nucleation (and hence induction time) reduces at higher supersaturation and temperature.

It should be noted that these relationships were derived from the data of He *et al* for 1M NaCl solutions, and thus the induction times are much shorter due to the catalytic effect of salt on induction time. This effect is observed by comparing the induction time of the 90% RO concentrate and the 1M NaCl solution from He *et al* at the same supersaturation *i.e.* $S_r = 5$ in Figure 3.5. At 90% recovery (0.1M) the induction time was 330 minutes. Whereas, in the 1M NaCl solution from He *et al,* at ten times the ionic strength of the 90% concentrate, the induction time was 9 times shorter (36 minutes).

In addition, the RO concentrate also contain particles which may increase induction time which are absent from the 1M NaCl solutions from He *et al*. However, despite these differences the factor by which the induction time increases for a temperature change of 20°C at a chosen supersaturation for the 1M NaCl solutions could be applied to correct the RO induction time as a first estimate. Thus, for a surface water RO concentrate with an average supersaturation ratio of 5 a temperature decrease of 20°C (assuming the supersaturation ratio was constant) would increase the induction time by a factor of 3.6.

3.4.2   Growth

3.4.2.1   Effect of Organic Matter

To establish if inhibition of $BaSO_4$ crystal growth by organic matter also contributes to the stable supersaturation observed in the RO system, the *initial linear growth rate* of RO and blank synthetic concentrate were determined. The *initial linear growth rate*s for the concentrates reported in Table 3.5 were determined from the slope of the best fit line for the desupersaturation data for the first hour, as this is taken as representative of the high initial supersaturation. Figure 3.6 presents an example for the 90% RO and blank synthetic concentrate.

The *initial linear growth rate* determined for RO concentrates at 80 and 90% recovery were lower than the blank synthetic concentrate (Table 3.5). This difference can be attributed to adsorption of organic matter present in the RO concentrate onto the seed crystal surface at growth sites such as crystal steps which inhibits growth resulting in a slower decrease in barium depletion [1,19]. The organic matter appears to be more effective in suppressing the growth rate at the higher recovery of 90%. The growth rate in the synthetic concentrate (3.0×10^{-13} m/s) is more than double that measured in the RO concentrate (1.2×10^{-13} m/s) at 90% recovery. Whereas, the same comparison at 80% recovery shows that the growth rate in the synthetic concentrate (1.6×10^{-13} m/s) is only a factor 1.2 higher than that of the RO concentrate (1.3×10^{-13} m/s). However, the growth rate measured in the 80% RO concentrate was higher than expected, being similar to that of the 90% RO concentrate. A lower growth rate was expected due to the lower supersaturation in the 80% concentrate. This is most likely due to variations in the organic matter over the sampling period.

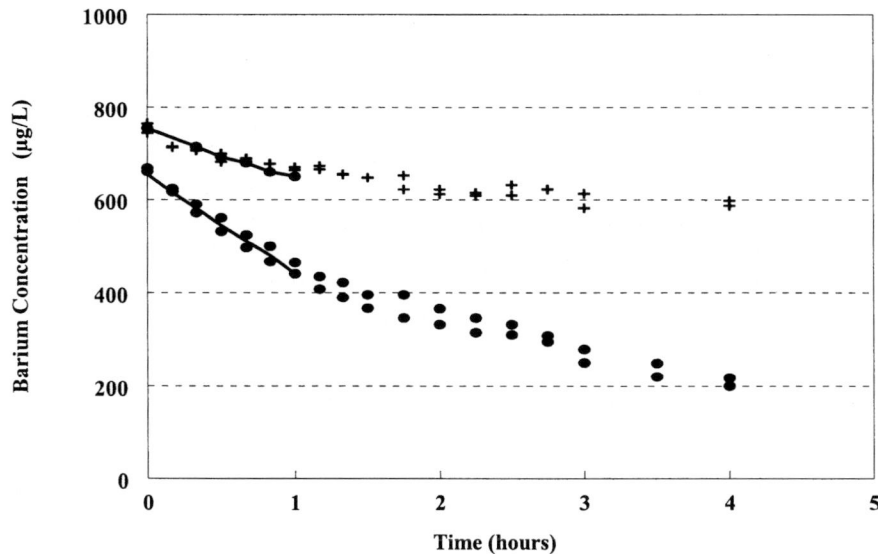

Figure 3.6: Barium concentration over time in duplicate seeded growth experiments for RO (10mg/L DOC) (✚) and blank synthetic concentrate (no organic matter present ≤0.1mg/L DOC) (●) at 90% recovery. Determination of the initial linear growth rate from the best fit straight line for the first hour of the desupersaturation data.

Table 3.5: The initial linear growth rate of barium sulfate in RO concentrate and synthetic concentrate; containing no organic matter (≤0.1mg DOC/L) and with humic acid (HA) addition determined at 25°C for 80 and 90% recovery with supersaturation expressed as growth affinity.

Recovery (%)	DOC Concentration (mg/L)	Initial Linear Growth Rate, G_{lin} (10^{-13} m/s)	Rate Constant, K_g (10^{-14} m/s)	Growth Affinity, β
80	RO (5mg/L)	1.3	3.1	2.22
	blank (≤0.1mg/L)	1.6	4.9	2.05
	HA (5mg/L)	0.7	2.2	
	HA (10 mg/L)	0.6	1.8	
90	RO (10mg/L)	1.2	0.8	3.19
	blank(≤0.1mg/L)	3.0	2.8	2.89
	HA (10mg/L)	0.5	0.5	
	HA (20mg/L)	0.9	0.9	

To observe the inhibition of crystal growth over time the linear growth rate was plotted *vs* growth affinity in Figure 3.7 for a 90% RO and a blank synthetic concentrate. In both cases initially high growth rates were observed which gradually slowed down as the concentration of barium decreased. The linear growth rate of the blank synthetic concentrate was observed to be significantly higher than the RO concentrate when compared at the same growth affinity. Nevertheless, despite the slower barium sulphate growth rates in the 80 and 90% RO concentrate, the barium decreased to the same solubility concentration as the blank synthetic concentrate. This can be explained by *overgrowth* of the inhibiting compound. Although, some of the growth sites are blocked by adsorbed organic matter, some of the growth sites may still be free to grow [19]. These slower advancing steps overgrow the organic matter, incorporating them into the crystal lattice and growth resumes but at a lower rate than the blank.

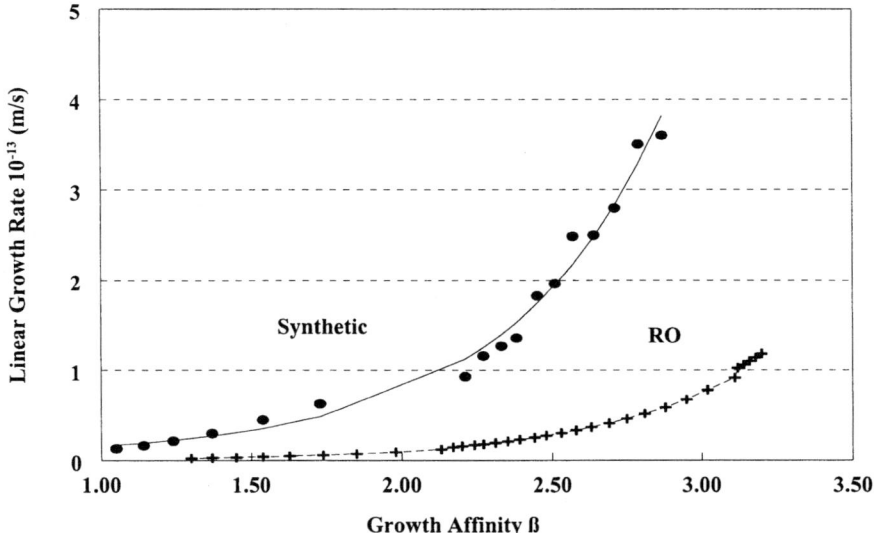

Figure 3.7:   Barium sulfate linear growth rate at 25 °C as a function of growth affinity in a RO (10mg/L DOC) (✚) and a blank synthetic concentrate (no organic matter present ≤0.1mg/L DOC) (●) at 90% recovery.

The effect of humic acid addition on the linear growth rate at 80 and 90% recovery in synthetic concentrates is presented in Figure 3.8 A and B, respectively. In this case, different linear growth rate profiles were found, where the linear growth rate decreased dramatically within the first hour in comparison to the respective blank. The *initial linear growth rates* and constants determined for the 80 and 90% synthetic concentrates in the presence of humic acid are included in Table 3.5.

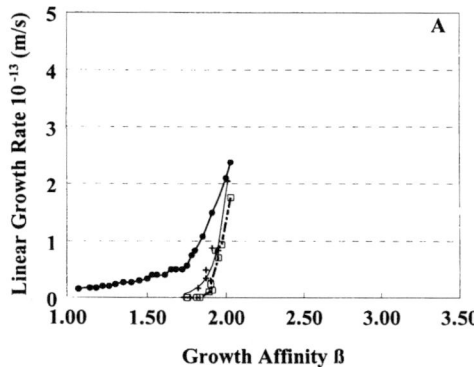
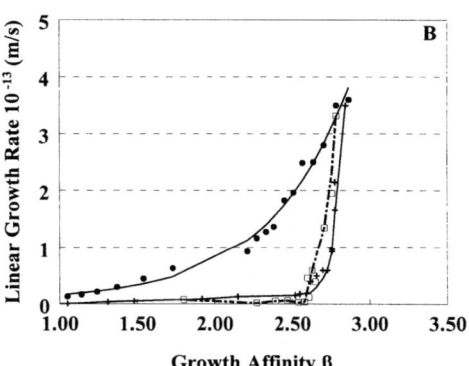

Figure 3.8: Barium sulfate linear growth rate at 25°C as a function of growth affinity in synthetic concentrate at 80% (A) and 90% (B) recovery containing; no organic matter ≤0.1mg/L DOC (●), 5mg/L and 10mg/L humic acid DOC at 80% and 90% recovery, respectively (✚), and 10mg/L and 20mg/L humic acid DOC at 80% and 90% recovery, respectively (□).

For the 80% synthetic concentrate with humic acid addition, the *initial linear growth rate* appeared to be dose independent. The initial linear growth rate in the blank synthetic concentrate was 1.6×10^{-13} m/s, and with the addition of 5mg/L and 10 mg/L humic acid DOC, this was reduced by more than half to 0.7×10^{-13} m/s and 0.6×10^{-13} m/s, respectively. Subsequently, the linear growth rate decreased to zero in the concentrate with humic acid addition (refer Figure 3.8A) indicating complete growth inhibition at a β of 1.75 which corresponded to 25 hours. At this point the barium concentration remained stable for both concentrate for the entire experiment duration of 200 hours.

In Figure 3.8B, the linear growth rate of the 90% concentrate containing humic acid significantly decreased within 10 minutes. Correspondingly, the *initial linear growth rates* in Table 3.5 were significantly lower, 3-6 times, in comparison to the blank synthetic concentrate (3.0×10^{-13} m/s). As in the 80 and 90% RO concentrates, overgrowth of the adsorbed inhibiting organic matter was evident in the 90% synthetic concentrate with humic acid. A comparison of the linear growth rate in Figure 3.8B for the 90% synthetic concentrate with 10 and 20mg/L humic acid DOC to the 90% RO concentrates with 10mg/l DOC (refer Figure 3.7) shows that the growth rate was never totally inhibited.

Overgrowth of the adsorbed humic acid is more easily observed in Figure 3.9. The barium concentration decreased to equilibrium solubility within 200 hours for the concentrate with 10mg/L humic acid DOC, albeit at a much slower rate than the blank synthetic concentrate (50 hours). Whereas, the synthetic 90% concentrate with the higher amount of humic acid (20mg/L) did not reach equilibrium solubility within the experiment duration. At the end of the experiment, the growth affinity was 1.8 and the linear growth rate was very low 8×10^{-15} m/s (refer Figure 3.8B) and eventually it is expected to reach equilibrium.

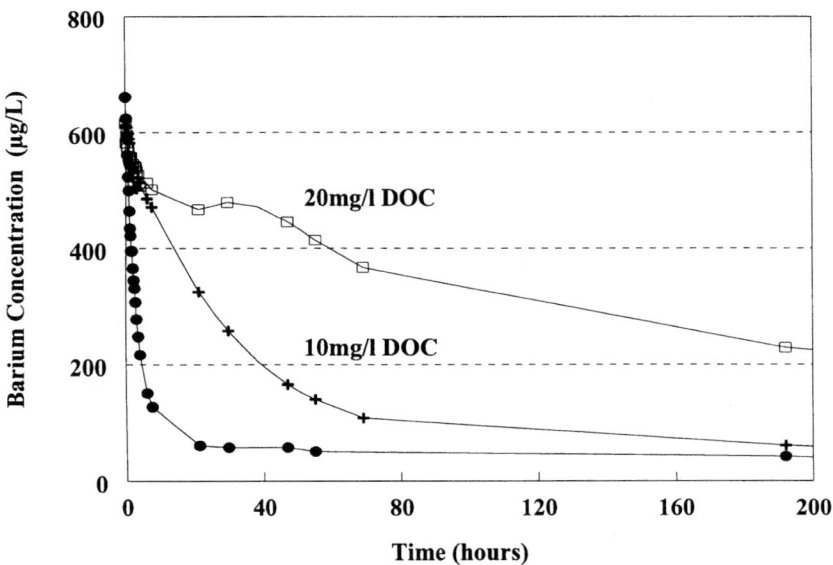

Figure 3.9:   Barium concentration over time in seeded growth experiments for synthetic concentrate at 90% recovery containing; no organic matter ≤0.1mg/L DOC (●), 10mg/L humic acid DOC (✚), and 20mg/L humic acid DOC (□).

From Table 3.5 the organic matter present in the RO feedwater (5mg/L DOC) at 80% recovery reduced the initial growth rate in comparison to the blank concentrate only by a factor of 1.2. Whereas, in the presence of humic acid the growth rate reduced by up to a factor of 2.7. Barium sulphate growth was totally inhibited in the 80% concentrate with humic acid but the organic matter in the AWS did not give total growth inhibition at 80% recovery.

At the higher recovery of 90% the initial linear growth rate was reduced by a factor of 2.5 by the organic matter in the RO feedwater and up to a factor of 6 in the presence of humic acid in synthetic concentrate. However, the humic acid could not completely inhibit barium sulphate crystal growth and the higher supersaturation overrides the inhibition.

Based on these results, RO systems treating surface water which contains a high concentration of humic acid may expect inhibition of barium sulphate precipitation at lower supersaturation. However, as supersaturation increases the chance of overgrowth of the adsorbed organic matter may occur resulting in barium sulphate precipitation. In the case of RO concentrate the barium sulphate growth rate is lowered by the presence of organic matter but not enough to result in stable supersaturation especially if barium sulphate crystals are attached to the membrane surface.

3.4.2.2   Effect of Supersaturation

From the previous section the initial linear growth rate determined at 80 and 90% recovery were similar despite the difference in supersaturation. Therefore, the results of the linear growth rates calculated for RO concentrate ranging from 86-90% recovery are presented as a function of growth affinity (refer Figure 3.10). The linear growth rates obtained fall onto a single curve with the 90% recovery concentrate (highest S_r) giving the highest initial growth rate. As the concentrates were collected on the same day the organic matter was constant and only supersaturation varied with recovery. The fact that a common curve was found in Figure 3.10 confirms that as expected from literature, growth affinity (and hence supersaturation) plays a major role in determining the growth rate [23].

The initial linear growth rate measured in the RO concentrate was significantly lower, by a factor of 60-120 than those reported in literature for pure water solutions (76-146 $\times 10^{-13}$ m/s) of a similar supersaturation at 25°C given in Table 3.6 [1, 25]. Likewise, the rate constant for the RO concentrate was also significantly lower than the pure water solutions. This is due in part to the organic matter present in the concentrate as discussed in the previous section. A comparison of the initial linear growth rate measured in the blank synthetic concentrate at 90% recovery (3.0 $\times 10^{-13}$ m/s) is still a factor 18 to 35 lower than those reported for pure water (refer Table 3.5 and 6).

Table 3.6:   Linear growth rates and rate constants reported in literature at 25°C compared to that determined for the 90 % RO concentrate with supersaturation expressed as growth affinity β.

Linear Growth Rate G_{lin} (10^{-13} m/s)	Rate Constant K_g (10^{-14} m/s)	β	Source
147	41.7	3.88	Liu et al [25]
76	47.4	3.2	van der Leeden [1]
1.2	0.8	3.19	[this study]

This inhibitory effect may be due to other factors such as the SO_4^{2-} to Ba^{2+} ion stoichiometry and the higher ionic strength in the concentrate. The 1000:1 SO_4^{2-} to Ba^{2+} ion stoichiometry in the RO concentrate may give rise to a low barium ion impingement at the crystal surface reducing the growth rate. In addition, the higher ionic strength of the RO concentrate in comparison to the pure water solutions reported in Table 3.6 could retard $BaSO_4$ growth. This is reflected in the decreasing rate constants in all the concentrates when the recovery was increased from 80 to 90% (refer Table 3.5). For example the rate constant of the blank synthetic concentrate decreased from 4.9 $\times 10^{-14}$ m/s to 2.8 $\times 10^{-14}$ m/s as the recovery was increased from 80 to 90%.

Nevertheless, despite the lower growth rates in the RO concentrate, the growth rate is significant at both 80 and 90% recovery. Notwithstanding this no scaling was found in the RO system. Thus, it is likely that growth is not limiting barium sulphate precipitation in the RO system.

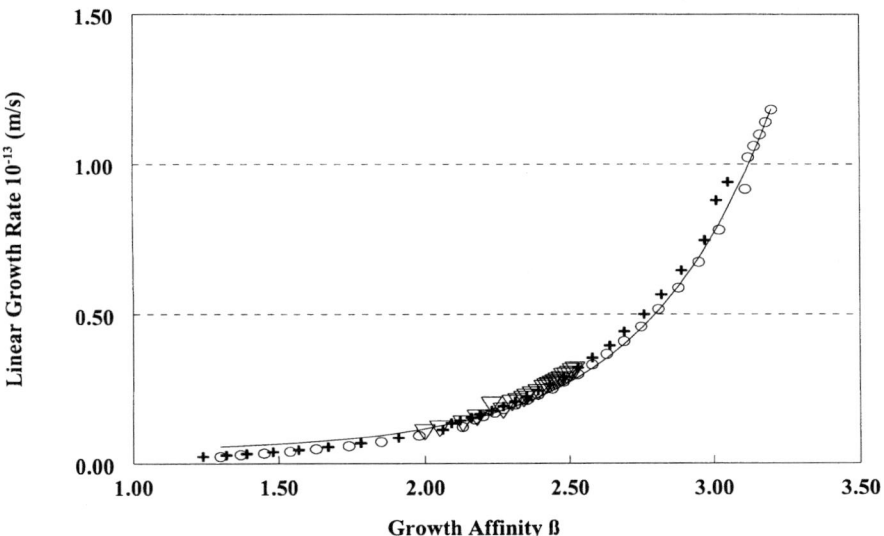

Figure 3.10:   Barium sulfate linear growth rate at 25°C as a function of growth affinity in RO concentrate at different initial supersaturation ratio: 4.94 (90% Recovery) (O); 4.64 (89% Recovery) (+); 3.52 (86% Recovery) (∇).

3.5   Conclusions

Supersaturation (expressed as S_r) was found to play a major role in determining the barium sulphate induction time and hence the stability of the RO concentrate. A 10% increase in recovery from 80% ($S_r = 3.1$) to 90% ($S_r = 4.9$) decreased the induction time from >200 to 5.5 hours, respectively. Temperature was found to have a significant effect on induction time using data published by He *et al* [17] and is most pronounced at lower supersaturation. A temperature decrease of 20°C from 25 to 5°C is expected to extend induction time by a factor of 3.6 for a supersaturation ratio of 5.

Organic matter present in the RO concentrate did not inhibit nucleation. Whereas, addition of commercial humic acid (20mg/L) to the 90% synthetic concentrate extended the induction time by more than 50 times. From the results of the growth experiments it was concluded that this was most likely due to inhibition of growth to a detectable size in the induction time measurement.

The barium sulphate growth rate in the RO concentrate was found to be 60-120 times lower than those reported in literature for pure water solutions of a similar supersaturation. The growth rate generally increased with increasing supersaturation. Organic matter in the 90% RO concentrate retarded the initial growth rate and was a factor 2.5 lower than in synthetic concentrate. However, growth was not totally inhibited and overgrowth of the adsorbed organic matter allowed barium desupersaturation. Commercial humic acid was more effective and total inhibition of growth in synthetic 80% concentrate after 25 hours was observed. Although, at higher supersaturation, growth inhibition was overridden and overgrowth of humic acid occurred leading to desupersaturation in the synthetic 90% concentrate.

Inhibition of nucleation by organic matter present in the RO concentrate does not limit barium sulphate precipitation. The induction time at 80% recovery was found to be very long (>200hours) which most likely prevents scaling in the RO system. Whereas, the growth rates measured for 80 and 90% RO concentrate were significant. If the growth rate was limiting precipitation then scaling should have occurred at 80% recovery in the pilot plant especially over the prolonged period of operation at this recovery. However, this was not the case and thus, growth is not the factor governing scaling. Results from this study indicate stable supersaturation in the RO concentrate is attributed to long induction times (slow nucleation kinetics).

Symbols

A	pre-exponential factor (nuclei/cm^{3}s)
A_1	constant in Equation 3.6 (1/s)
A_2	constant in Equation 3.12
$A_{(0)}$	initial surface area of seed crystals (m^2)
$A_{(t)}$	total surface area of seed crystals at time t (m^2)
B	defined by Equation 3.5 (-)
dc/dt	concentration differential (mol/Ls)
c	concentration (mol/L)
E	constant in Equation 3.4
E_a	activation energy (J/mol)
F	geometrical shape factor (for spherical nucleus 16π/3)
ΔG_{cr}	free energy change for the critical cluster size (J/molecule)
G_{lin}	mean linear growth rate (m/s)
J_n	nucleation rate (nuclei/cm^{3}s)
k	Boltzmann constant (J/K molecule)
K_g	temperature dependent growth rate constant (m/s)
K_{sp}	BaSO$_4$ thermodynamic solubility product (mol/L)^2
M_{BaSO4}	molar mass of barium sulphate (kg/mol)
M	Molal (mol/L water)
$M_{(0)}$	initial mass of seed crystals (kg)
$M_{(t)}$	total mass of seed crystals at time t (kg)
N_a	Avogadro's number (molecules per mol)
p	order of the growth rate (-)
R	universal gas constant (J/K mol)
S_r	supersaturation ratio (-)
t_{ind}	induction time (s)
T	absolute temperature (K)
dV/dt	crystal volume differential (m^3/s)
V_s	volume of the solution (m^3)
V_m	molar volume of solid (cm^3/mol)
[]	concentration (mol/L)

Greek Symbols

β	growth affinity (-)
γ_+	ionic activity coefficient of cation (-)
γ_-	ionic activity coefficient of anion (-)
$f(\theta)$	contact angle factor (-)
Δμ	difference in chemical potential of one mole electrolyte in its crystalline form and in its supersaturated solution (J/mol)
ρ	density of barium sulphate crystals (kg/m³)
σ	interfacial tension (J/m²)

Abbreviations

BET	Brunauer Emmet Teller model
DOC	Dissolved Organic Carbon
LSI	Langelier Saturation Index

References

1.	M.C. Van der Leeden, The Role of Polyelectrolytes in Barium Sulfate Precipitation, Ph.D Thesis, TU Delft, 1991.
2.	Permasep Engineering Manual (PEM), Du Pont de Nemours & Co. 1982.
3.	J.P. Van der Hoek, P.A.C. Bonné, E.A.M. Van Soest and A. Graveland, Application of hyperfiltration at the Amsterdam Waterworks: Effect of pretreatment on operation and performance. Proceedings of the American Waterworks Association 1995 Membrane Technology Conference, August 13-16 Reno, U.S.A. 277-294.
4.	J.P. Van der Hoek, P.A.C. Bonné, and E.A.M. Van Soest, Fouling of reverse osmosis membranes: The effect of pretreatment and operating conditions. Proceedings of the American Waterworks Association 1997 Membrane Technology Conference, February 23-26, New Orleans, U.S.A.
5.	S.F.E. Boerlage, M.D. Kennedy, G.J. Witkamp, J.P. Van der Hoek and J.C. Schippers, BaSO₄ solubility in reverse osmosis concentrates, J. Mem. Sci., 159 (1999) 47-59.
6.	J.P. Van der Hoek, P.A.C. Bonné and E.A.M. Van Soest, Reverse osmosis: Finding the balance between fouling and scaling, Proceedings of the 21st IWSA World Congress 1997, September 20-26, Madrid, Spain.
7.	C.C.Templeton, Solubility of barium sulphate in sodium chloride from 25° to 95°C. J. of Chem. Eng. Data, 5 (1960) 514-516.
8.	A.S.Myerson, Handbook of Industrial Crystallization. Butterworth-Heinemann, 1993.
9.	J.W. Mullin, Crystallization, 3rd Edition, Butterworth-Heinemann, 1993.
10.	J.I. Drever, The Geochemistry of Natural Waters. 2nd Edition. Prentice Hall Englewood Cliffs, New Jersey 07632, USA 1988.
11.	E.M. Thurman, and R.L. Malcolm, Preparative isolation of aquatic humic substances, Env. Sci. Technol. 15 (1981) 4, 463-466.
12.	J.C. Cowan, Water-Formed Scale Deposits, Gulf Publishing Company, 1992.
13.	O. Söhnel and J. Garside, Precipitation - Basic Principles and Industrial Applications, Butterworth-Heinemann, 1992.
14.	A.E. Nielsen, Kinetics of Precipitation Pergamon Press, New York, 1967.
15.	O. Söhnel and J.W. Mullin, Interpretation of crystallization induction periods, J of Coll. Inter. Sci., 123 (1988) 43 - 50.
16.	J.W. Zhang and G.H. Nancollas, Mechanisms of growth and dissolution of sparingly soluble salts. in: Reviews in Mineralogy, 23 (1990) 365 - 396.

17.   S. He, J.E.Oddo and M.B.Tomson, The nucleation kinetics of barium sulfate in NaCl solutions up to 6M and 90°C, J. of Coll. and Inter. Sci, 174 (1995) 319-326.

18.   D.J. Gunn, and M.S. Murthy, Kinetics and mechanisms of precipitations, Chem. Eng. Sci. 27 (1972) 1293-1313.

19.   S.T. Liu, and G.H. Nancollas, A kinetic and morphological study of the seeded growth of calcium sulphate dihydrate in the presence of additives, J. Coll. and Inter. Sci., 52 (1975) 593-601.

20.   S. He, J.E. Oddo, and M.B. Tomson, The nucleation kinetics of calcium sulfate dihydrate in NaCl solutions up to 6M and 90°C, J. of Coll. and Inter. Sci, 162 (1994) 297-303.

21.   G.H. Nancollas, and N. Purdie, Crystallization of barium sulphate in aqueous solution, Trans. Far. Soc., 59 (1963) 735-740.

22.   G.M. Van Rosmalen, M.C. van der Leeden and J. Gouman, The growth kinetics of barium sulphate crystals in suspension: Scale prevention (I), Kristall und Technik, 15, (1980), 1213-1222.

23.   G.M. Van Rosmalen, P.J. Daudey and W.G.J. Marchee, An analysis of growth experiments of gypsum crystals in suspension, J. Crystal Growth, 52 (1981), 801-811.

24.   A.E. Nielsen and J.M. Toft, Electrolyte crystal growth kinetics, J. Crys. Growth, 67 (1984) 278-288.

25.   S.T. Liu and G.H. Nancollas, The crystal growth and dissolution of barium sulphate in the presence of additives, J. Coll. and Inter. Sci., 52 (1975) 582-592.

26.   S.T. Liu, G.H. Nancollas and E.A. Gasiecki, Scanning electron microscopic and kinetics studies of the crystallization and dissolution of barium sulphate crystals, J. Crys. Growth, 33 (1976) 11-20.

27.   F.Van der Ham, Pitzer Model from Laboratory for Process Equipment, Delft University of Technology, Leeghwaterstraat 44, 2628 CA Delft, The Netherlands

28.   S. He, J.E. Oddo, and M.B. Tomson, The inhibition of gypsum and barite nucleation in NaCl brines at temperatures from 25 to 90°C. App. Geochemistry, 9, (1994) 561-567

29.   B.D. Symeopoulos and P.G. Koutsoukos, Spontaneous precipitation of barium sulfate in aqueous solution. J. Chem. Soc. Faraday Trans., 88 (1992) 3063-3066.

30.   R.L. Malcolm and P. MacCarthy, Limitations in the use of commercial humic acids in water and soil research. Environ. Sci. Technol. 20 (1986) 904-911.

31.   J. G. Hering and F.M.M. Morel, Humic acid complexation of calcium and copper. Env. Sci. Tech. 22 (1988) 1234-1237.

32.   H. Schubert, and A. Mersmann, Determination of heterogenous nucleation rates, Trans. I. Chem. E. 74 (1996) 821-827.

33.   G.H. Nancollas, The growth of crystals in Solution, Adv. Colloid Interface Sci, 10 (1979) 215-252.

34.   M.C. van der Leeden, D. Kashchiev and G.M. van Rosmalen, Precipitation of barium sulfate: Induction time and the effect of an additive on nucleation and growth. J. Coll. and Inter. Sci, 154, (1992) 338-350.

4

Scaling Prediction in Reverse Osmosis Systems: Barium Sulphate

Chapter 4 is based on: "Scaling Prediction in Reverse Osmosis Systems: Reverse Osmosis" by S.F.E. Boerlage, I. Bremere, M.D. Kennedy, G.J. Witkamp, J.P. Van der Hoek and J.C. Schippers. Submitted to *Journal of Membrane Science*.

Contents

Abstract

Barium sulphate scaling in reverse osmosis (RO) causes flux decline and potentially severe membrane damage. Existing methods e.g. Du Pont, predict barite scaling when the RO concentrate is saturated. However, precipitation in supersaturated concentrate may be limited by slow nucleation kinetics, expressed as long induction times. Induction time decreases, hence precipitation is more likely, with supersaturation and temperature. This research aims to develop a more realistic method to predict barite scaling based on the assumption that a threshold induction time can be defined which should not be exceeded to prevent scaling. Induction times were calculated for supersaturation (determined using the Pitzer model) and temperature data from an RO pilot plant from a relationship derived from measured induction times at 25°C. Safe (≥ 10 hours) and unsafe (≤ 5 hours) induction time limits, were derived from periods when scaling did and did not occur in the RO system at recoveries between 86-90%. Based on these induction times, safe and unsafe supersaturation limits were defined for 5-25°C. Use of these limits allows more flexible operation in optimising RO recovery while avoiding scaling. The general validity of these limits should be verified in further pilot studies with feedwater of different quality and using different RO elements.

4.1   Introduction

Inorganic scale formation on reverse osmosis (RO) and nanofiltration (NF) membranes is widely acknowledged as a potential problem when applying these technologies to surface, ground and sea water sources. Scaling is the result of the precipitation of sparingly soluble salts within the membrane module, e.g. on the membrane surface and/or spacer of a spiral wound element or the (non)-woven fabric in a Du Pont hollow fibre permeator. In practice, first a decrease in permeability is observed, resulting in a higher feed pressure to maintain productivity and recovery. Subsequently, an increase in differential pressure across a membrane element/module is observed (pressure drop in the feed/brine channel, in a spiral wound element). As a consequence, additional energy is required and more frequent cleaning. Besides the well known sparingly soluble salts, calcium carbonate and calcium sulphate, barium sulfate might be especially problematic if present in RO and NF feedwater due to its extremely low solubility, 2.33mg/L in pure water [1]. Moreover, if not observed in time, barium sulphate (barite) scale may age into a hard adherent layer. Cleaning may not then be effective and membrane elements have to be replaced.

The ability to predict scaling is an important tool in its control, both at the design stage and for continuous monitoring during plant operation. The most common method for predicting scaling in membrane systems is based on predicting if the solubility of the salt will be exceeded at design recovery i.e. the solution is supersaturated. The method in the Du Pont Manual, which is the most widely applied method in the RO industry to predict barite scaling, is based on this principle [2]. However, in earlier research this method was found to underpredict solubility as it did not adequately correct for ionic strength effects in RO concentrate. Moreover, the Du Pont method is limited to solubility prediction at 25°C while barite solubility decreases with temperature [3].

If scaling is predicted typically two methods are employed to avoid its occurrence, either lowering the recovery below the solubility of the salt or by antiscalant and/or acid addition, the recovery can be increased to a certain extent. However, operating a plant at a higher recovery is attractive because less feedwater is required including pretreatment costs, the energy requirement per m^3 is lower and less concentrate needs to be disposed of. In the case of barium sulphate, lowering the recovery may not even be an option, if the feedwater is already close to saturation. Moreover, antiscalants have been reported to promote biofouling due to their designed biodegradability in order to result in more environmentally favourable concentrate disposal [4,5]. Therefore, newer antiscalants have been developed which although being biodegradable, do not give rise to biofouling.

This research is largely based on results from the Amsterdam Water Supply (AWS) RO pilot plant studies, where RO is being investigated as an integrated part of two treatment systems for pretreated River Rhine water. One pilot plant study objective was to investigate the conditions under which the RO system could be operated without barium sulphate scaling. As according to estimates using the Du Pont method, the feedwater was already supersaturated with barium sulphate and predicted to scale.

Based on this prediction the RO pilot plant was initially operated with antiscalant and sulfuric acid addition at 90% recovery to prevent barite and alkaline scaling, respectively. Nevertheless, significant barite scaling was found [6]. Changing the antiscalant prevented scaling, however biofouling occurred and therefore antiscalant addition was terminated [5-7]. The recovery was then lowered from 90% to 80%, and hydrochloric acid replaced sulfuric acid addition to reduce the sulphate ion concentration in the feedwater and the risk of barite scaling. This strategy was successful since no scaling was observed during nineteen months of operation. Subsequently, plant recovery was again increased stepwise, under the same conditions in order to establish to what extent the recovery could be increased safely. The result was that barium scaling occurred at 88-90% recovery [7]. The recovery was therefore again lowered to 85% and no scaling was observed for a further year of operation at this recovery [8].

These results clearly demonstrate that solubility based methods such as that in the Du Pont Manual greatly overestimate the threat of barium sulphate scaling. This is because precipitation kinetics, crucial for scale prediction, are neglected. Although, precipitation may be thermodynamically possible in a saturated solution, slow precipitation kinetics (nucleation followed by growth) may result in *metastability* of the solution [9-13]. Result from the previous chapter indicate that most likely crystal growth does not to limit precipitation in the RO concentrates [14]. Rather slow nucleation kinetics i.e. the formation of a new crystal surface, measured as long "induction times" was found to be the cause of stable supersaturation in the 80 and 85% RO concentrate. Where the probability of scaling increases as the induction time decreases [14].

Induction time is principally determined by supersaturation and temperature. Consequently, a recovery increase, generally shortens induction times and increases the risk of scaling as supersaturation increases. Whereas, changes in RO feedwater temperature have a double effect on barium sulphate induction time which act in opposition. A temperature increase, directly shortens the induction time but also decreases the supersaturation and hence extends the induction time [9-12]. Thus, induction time is the net effect of temperature changes and supersaturation.

The aim of this research is to develop a method which more realistically predicts under which supersaturation and temperature conditions, barium sulphate scaling occurs in the RO system. This method assumes that an induction time can be defined which should not be exceeded to prevent scaling in the RO system. Induction times will be calculated for different supersaturation and temperature conditions from measured induction times at $25\,^{\circ}C$. The supersaturation ratio calculated with the Pitzer method using an experimentally determined K_{sp} at $25\,^{\circ}C$ corrected for temperature will be compared to that calculated by the Du Pont method. Periods in which scaling was observed are correlated with the calculated induction times. From these data "safe" and "unsafe" induction times are derived and translated into safe and unsafe supersaturation limits.

4.2   Background

4.2.1   Solubility Prediction

4.2.1.1   Du Pont Manual

The method in the Du Pont Manual for barium sulphate scaling prediction is based on predicting barium solubility at 25°C using a *solubility product*, K_C [2]. The K_C expresses the dynamic equilibrium between the scalant crystal ions in solution and the crystal solid phase (unity) defined for barium sulphate as follows [10,11]:

$$K_C = [Ba^{2+}][SO_4^{2-}] \text{ (at equilibrium)} \qquad\qquad ...(4.1)$$

A further increase in barium or sulphate ions through another source *e.g.* sulphuric acid addition, will increase the supersaturation of the solution and the overall solubility will decrease through the *common ion effect* in accordance with the K_C and equilibria principles.

At ionic strengths > 0.001M, electrical interactions occur between ions resulting in an increase in solubility through the *ionic strength effect* [9,11]. Thus, using a fixed K_C, is no longer accurate. To account for ionic strength effects the Du Pont method corrects the K_C from a graphical relationship of K_C as a function of ionic strength for the desired recovery. The K_C is then compared with the ion product of the scalant ions (molarity). If the ion product exceeds K_C, the solution is supersaturated and scaling is predicted to occur [2].

To avoid scaling, the Du Pont method recommends antiscalant addition or a recovery which ensures that the scalant concentration product is 20% below the K_C [2]. This ensures a safety factor for concentration polarisation, which may occur at the membrane surface which causes a higher localized concentration than in the bulk solution.

4.2.1.2   Activity Method

A more accurate approach to predicting solubility in solutions of high ionic strength such as RO concentrate employs the *thermodynamic solubility product* K_{sp} which uses activities which take into account these ionic interactions [9-11]:

$$K_{sp} = a_{Ba^{2+}(aq)}\, a_{SO_4^{2-}(aq)} = \gamma_+ [Ba^{2+}]\gamma_- [SO_4^{2-}] \qquad\qquad ...(4.2)$$

where the activity $a_{i(aq)}$ is the product of ion i molal concentration and its *activity coefficient* γ_i, (dimensionless). The K_{sp} is dependent on temperature and pressure but is independent of ionic strength.

The $K_{sp25°C}$ (1.34×10^{-14}) for the RO concentrate in the AWS pilot plant was experimentally determined in Chapter 2 [15]. The K_{sp} (and barium solubility) increases with temperature.

Therefore, an empirical relationship was derived in Chapter 2 (section 2.4.5) from barium solubility data in pure water reported by Templeton [3] and Linke [16] for the K_{sp} as a function of temperature (range 5-25 °C) as follows:

$$K_{sp_T} = \left(\frac{T_{°C}}{25°C}\right)^{0.634} \times K_{sp_{25°C}} \qquad \qquad ...(4.3)$$

The thermodynamic K_{sp_T} can then be used to accurately quantify supersaturation in the RO concentrate for temperature T by calculating the supersaturation ratio S_r as follows [9-11]:

$$S_r = \sqrt{\frac{\gamma_+ [Ba^{2+}] \gamma_- [SO_4^{2-}]}{K_{sp_T}}} \qquad \qquad ...(4.4)$$

4.2.2   Kinetics

Below a supersaturation of 1, the solution is under-saturated and precipitation is not possible (refer stable zone in Figure 4.1). Thermodynamically, crystallization or precipitation is feasible once the solution is supersaturated at a given temperature. However, a certain supersaturation level is required before spontaneous crystallization occurs, represented by the broken line in the solubility-supersaturation diagram (Figure 4.1). Below this level solutions appear to be stable, referred to as *"metastable"* in literature [9-13]. This metastability is attributable to slow precipitation kinetics. The mechanism of precipitation comprises two kinetic processes; nucleation, *i.e.* "birth" of a crystal followed by crystal growth [9-11].

Normally, metastability can be attributed to a low nucleation rate. Inhibition of nucleation by the presence of impurities may also occur, however, this is rarely observed. Alternatively, subsequent inhibition of crystal growth by impurities such as organic matter in the solution may also result in "stability" being observed. In the case of the RO concentrate, the limiting step in precipitation kinetics was found in Chapter 3 to be more likely due to slow nucleation kinetics, measured as long induction times, rather than crystal growth inhibition.

4.2.2.1   Nucleation

In a supersaturated solution the crystal ions associate into clusters by an addition mechanism that continues until a critical size is reached. Below the critical size the probability that the cluster survives is limited, therefore, these clusters typically redissolve. Whereas, clusters above the critical size grow giving stable "nuclei", the new crystal surface. Primary nucleation occurs in the absence of the crystalline solid being formed. Two types of primary nucleation can be distinguished; *homogeneous* (spontaneous) and *heterogeneous*. Homogeneous is dominant at high supersaturation while heterogeneous, induced by foreign particles, the membrane surface or dissolved system impurities can occur at lower supersaturations.

A metastable zone in supersaturation exists where the nucleation rate is so low that precipitation is improbable within the measured time frame (middle zone in Figure 4.1) [9,11,17]. The width

of this zone is given thermodynamically by the free energy of formation of the critical cluster size in homogenous nucleation. Homogeneous nucleation forms the basis of several nucleation theories and can be extended to heterogeneous nucleation. Two important parameters characterize the nucleation process: (i) the free energy change, and (ii) the rate of nuclei formation. The nucleation rate, J_n (nuclei/cm³s), in classical theory is given by an Arrhenius type of expression [9,13]:

$$J_n = A \exp\left[\frac{-\Delta G_{cr}}{kT}\right] \qquad ...(4.5)$$

where A is the pre-exponential factor related to the efficiency of collisions of ions and molecules, which has a value of *circa.* 10^{23} to 10^{33}, k is the Boltzmann constant, T is the absolute temperature, and ΔG_{cr} is the free energy change for the critical cluster size to form. The nucleation rate can be expressed in molar quantities as well [12, 13]:

$$J_n = A \exp\left[-\frac{F V_m^2 \sigma^3 f(\theta) N_a}{(RT)^3 (\ln S_r^2)^2}\right] \qquad ...(4.6)$$

where F describes the crystal geometry (relating area and volume shape factors *e.g.* for a spherical nucleus $16\pi/3$), V_m is the molar volume of solid, σ the interfacial tension, $f(\theta)$ contact angle factor (unity for homogeneous nucleation), N_a is Avogadro's number, and R is the universal gas constant.

The nucleation rate is critically dependent on supersaturation and temperature. As supersaturation increases the critical cluster size decreases. Similarly, a temperature increase will increase the nucleation rate in Equation 4.6 when S_r is kept constant. Thus, when supersaturation and temperature are increased, nucleation becomes increasingly favourable and nucleation occurs spontaneously. This is referred to as the metastable limit (upper broken line in Figure 4.1). Above this limit, solutions are unstable and immediate crystallization is extremely likely [9,10].

The metastable limit is more difficult to define than the lower solubility boundary. For sparingly soluble salts such as barium sulphate the metastable zone may by broader and show a more marked temperature dependence than for readily soluble salts [12]. In the case of heterogeneous nucleation the presence of foreign particles or dissolved system impurities typically lowers the metastable limit. These impurities can act as a surface or nuclei catalysing nucleation by reducing the contact angle factor ($f(\theta)$ <1) and concomitantly lowering the free energy of formation of the critical cluster size [9-11]. RO systems constitute an impure system where impurities are present, and therefore heterogenous nucleation is most likely the dominant nucleation mechanism. Nucleation in the RO system can occur either in the RO bulk concentrate or at the surface of the membrane and/or spacer.

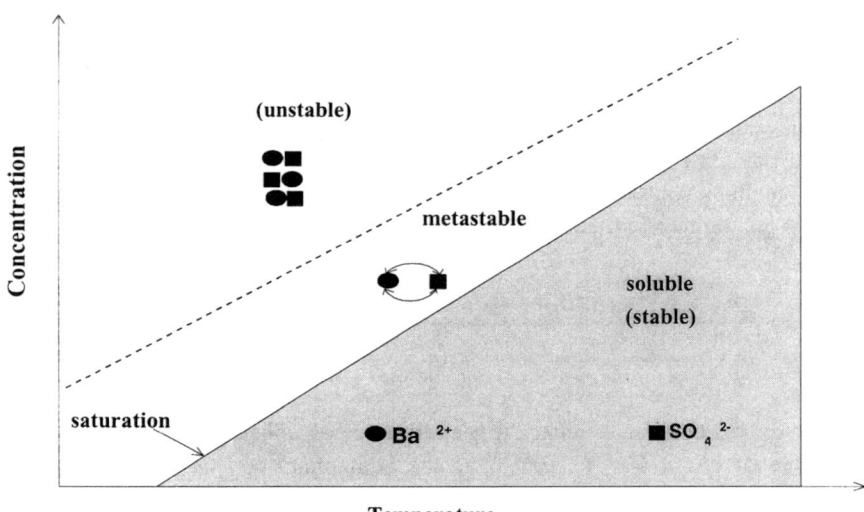

Figure 4.1:   Solubility-supersaturation zones for barium sulfate (i) undersaturated zone where barium sulphate is soluble and no precipitation is possible (ii) metastable (supersaturated) zone where spontaneous precipitation is not likely and (iii) unstable zone where spontaneous precipitation is extremely likely (adapted from Mullin [11]).

4.2.2.2   Induction Time

The induction time, t_{ind}, is frequently used to estimate nucleation time and is defined as the time elapsed between the creation of supersaturation and the first appearance of the new crystal solid phase, ideally nuclei with the critical cluster size dimensions [1,14,18]. However, as the induction time is determined experimentally, it may also include growth to a detectable size [1,19,20]. If it is assumed that the nucleation time is much greater than the time required for growth of crystal nuclei to a detectable size, then the induction time is inversely proportional to the rate of nucleation, $t_{ind} \propto J^{-1}_n$ and, Equation 4.6 can be written as [10,12]:

$$\log t_{ind} = \frac{B}{(\log S_r^2)^2} - E \qquad \qquad ...(4.7)$$

When a plot is made of $\log t_{ind}$ and $(\log S_r^2)^{-2}$ at a constant temperature, a linear region can be found where E is a constant and B the slope of the line is [12]:

$$B = \frac{F V_m^2 \sigma^3 f(\theta) N_a}{(RT)^3} \qquad \qquad ..(4.8)$$

However, two linear regions may be observed in a plot covering a wide range of supersaturation. The region of higher slope, visible at higher supersaturation, corresponds to homogeneous nucleation and at lower supersaturation a region of lower slope due to heterogeneous nucleation. The transition between these two regions may be smooth and connected by a continuous curve [10]. In the same plot, for a higher temperature the slope of the linear regions will decrease as induction time is shorter [10]. An empirical Arrhenius type relationship between temperature and induction time at a constant supersaturation was suggested by Gunn for $BaSO_4$ [21]:

$$\log \frac{1}{t_{ind}} = A_1 - \frac{E_a}{2.303\ RT} \qquad \qquad ...(4.9)$$

where E_a is the activation energy for nucleation and A_1 is a constant. Linearity was found by Gunn for a $BaSO_4$ supersaturation ratio of 20 and 29 when the temperature varied between 17-50°C. Induction time data available in the work of He *et al.* [18] for 1 M NaCl solutions with a $BaSO_4$ supersaturation ratio ranging between 5 and 30 for temperatures between 25-90°C also gave linearity using this relationship when tested in Chapter 3 (section 3.4.1.4).

4.3 Methods

4.3.1 Calculation of Supersaturation in RO concentrate

Barium and sulphate concentrations at the desired recovery (Y) were calculated from the respective concentrations measured in the RO feedwater multiplied by the corresponding concentration factor (CF), assuming 100% membrane rejection. To account for concentration polarisation the CF factor is multiplied by the concentration polarisation factor. The latter was estimated to be minimal for AWS, 1.01 [22] and was therefore neglected.

$$CF = \frac{1}{1 - Y} \qquad \qquad ...(4.10)$$

These concentrations were then input into Equation 4.4 to calculate the supersaturation ratio of the RO concentrate at the corresponding temperature measured for the RO feedwater by the activity method. Activity coefficients were calculated by the Blue Sea programme [23] which uses Pitzer equations as described in Chapter 2. However, as demonstrated in Chapter 2 the Bromley correlation could equally have been employed. The thermodynamic K_{spT} for temperature T (°C) was calculated using Equation 4.3.

In order to compare supersaturation calculated by the activity method and by the Du Pont method, Equation 4.11 was used. The supersaturation ratio for Du Pont (S_d) is the ion product of the barium and sulphate concentrations in the RO concentrate (calculated as above) divided by the Du Pont K_c:

$$S_d = \sqrt{\frac{[Ba^{2+}]\ [SO_4^{2-}]}{K_c}} \qquad \qquad ...(4.11)$$

4.3.2   RO Pilot Plant

The two reverse osmosis pilot installations at AWS, RO I and II, both use conventionally pretreated River Rhine water i.e coagulation, sedimentation and rapid sand filtration. In RO I, the RO feedwater undergoes further treatment by ozonation, biological activated carbon filtration and slow sand filtration prior to reverse osmosis. Whereas, RO II employs only an additional slow sand filtration step before reverse osmosis [6].

The reverse osmosis systems, following the pretreatment schemes, are identical. Each RO pilot installation is comprised of seven pressure vessels in a three stage Christmas tree configuration (4-2-1). Each pressure vessel contains six spiral wound elements. Toray membranes type SU 710L were used from November 1993- March 1997, afterwhich they were replaced by Fluid Systems, type 4821 ULP. The RO systems operate under constant flux using a frequency controlled high pressure feed pump and a PLC controlled concentrate valve. At 80% and 90% recovery 7.3 (flux 25 l/m²/hr) and 9.1m³/h (flux 31 l/m²/hr) of water is produced, respectively.

The various operating modes employed at the AWS plant are summarised in Table 4.1. Acid was added to prevent CaCO$_3$ scaling. More acid is necessary in the absence of an antiscalant as a lower LSI value in the concentrate is required. In the presence of an antiscalant, a higher LSI can be tolerated and acid was added according to the suppliers LSI guidelines i.e. an LSI of 1.5 and 2.0 for antiscalant A and B, respectively [6].

In this research, data from the RO II system were used. The RO system recovery was calculated on a daily basis from the RO permeate flow (Q_p) produced over the entire RO installation divided by the feedwater flow (Q_f). Barium sulphate deposition BA (mg/hr) data over the entire RO installation was supplied by AWS from a mass balance calculation using measured barium concentration C (mg/m³) and flow Q (m³/hr) in the feedwater $_{(f)}$, permeate $_{(p)}$ and concentrate $_{(b)}$ of the third stage as follows:

$$BA = C_f Q_f - C_b Q_b - C_p Q_p \qquad \qquad ...(4.12)$$

Scaling in this research was defined as barium deposition above 200mg/hr, due to the accuracy of the flow measurement system which gave a background "noise".

4.3.3   Calculation of Induction Time in the RO concentrate at 25°C

Induction time in the bulk RO I concentrate were measured in Chapter 3 in batch test experiments at 25°C. A relationship of induction time with supersaturation was determined for the RO concentrate (and extrapolated to higher supersaturation) from the experimental data (refer Chapter 3 section 3.4.1.2):

$$\frac{1}{\log t_{ind}} = 0.182 + 0.114 \left[\ln \frac{1}{(\log S_r^{\,2})^2} \right]^2 \qquad ...(4.13)$$

This function (valid for $S_r \geq 3.16$,) was used to calculate the induction time for RO concentrate and is assumed to be valid for either RO I and RO II concentrate.

Table 4.1: Operating modes at the AWS reverse osmosis pilot plants.

Operating Mode	Time Interval	Recovery (%)	Acid	LSI concentrate
1. Antiscalant A dose: 2-4mg/l Mol Wt. 1600	Nov 93 -May 94	90	H_2SO_4	1.5
2. Antiscalant B dose: 2mg/l Mol Wt. 3500	May 94 -Nov 94	90	H_2SO_4	2.0
3. No Antiscalant	Nov 94- July 96	80	HCl	0.0
4. No Antiscalant	9 July 96-23 Aug 86% 23 Aug -11 Oct- 88% 11 Oct -15 Nov 90%	86 88 90	HCl	0.0
5. No Antiscalant	April 97- March 98	85	HCl	0.0

4.3.3.1 Correction of Induction Time for Temperature

The calculated induction time at 25°C for the RO concentrate was multiplied by a temperature correction factor to estimate the induction time at the temperature T (Kelvin) of the RO feedwater. The temperature correction factor was calculated by determining the ratio of the induction time at temperature T (Equation 4.14) to that of 25°C (Equation 4.14), using the slope and intercept reported earlier (refer Chapter 3 section 3.4.1.4) from He *et al.* [18] (Equation 4.9) for 1M NaCl solutions. The slope and intercept employed were selected from Table 4.2 corresponding to the closest supersaturation value of the RO concentrate.

$$\log \frac{1}{t_{ind}} = \frac{1000}{T} A_3 + A_1 \qquad \qquad ...(4.14)$$

Table 4.2:   Slope and intercept obtained by Boerlage et al from plotting induction time data for $BaSO_4$ in 1M NaCl solutions reported by He *et al* [18] as the log reciprocal induction time *vs* temperature relationship for a given supersaturation (Equation 4.9).

Supersaturation Ratio S_r	Slope A_3	Intercept A_1
5	-2.31	4.41
10	-2.16	5.23
15	-1.66	4.05
20	-1.34	3.28

4.4   Results and Discussion

4.4.1   Operation at 90% recovery

4.4.1.1   Comparison of Methods to Calculate Supersaturation

The variation in supersaturation, calculated by the Du Pont and activity methods during operation at 90% recovery (mode 1 and 2) is presented in Figure 4.2. In Chapter 2 the Du Pont method, which is limited to solubility prediction at 25°C, was found to under predict solubility by between 30-40% at this temperature. Whereas, barium solubility predicted by the activity method was found to be accurate for the 5-25°C range and is presumed accurate for the entire temperature range shown in Figure 4.2. Consequently, the supersaturation calculated by the Du Pont method was significantly *overestimated* (>15%) by that calculated by the activity method when the RO feedwater temperature reached 25°C during mode 2.

The median RO feedwater temperature was significantly lower than 25°C during both modes. Mode 1 gave the lowest median temperature (6.5°C) as this operated during autumn and winter, while mode 2, operated mainly in the spring-summer period and hence the median temperature was higher at 16.6°C. Thus, the discrepancy in the supersaturation calculated by the Du Pont method generally increases with decreasing temperature and the real supersaturation in the RO concentrate is *underestimated*. A maximum difference of ≈-40% was found in mode 1 corresponding to the lowest RO feedwater temperature of 3.3°C. Supersaturation calculated for this temperature by the Du Pont method was 10.8 and by the more accurate activity method 17.5.

As expected a difference in the median supersaturation (1.8) was found between the two methods in mode 1. However, for mode 2, the median supersaturation calculated by the two methods was very close (refer Table 4.3). This is purely coincidental, due to the combination of errors of (i) solubility under prediction at 25°C and (ii) the temperature limitation of Du Pont. As the calculation of induction time requires the accurate quantification of supersaturation, henceforth, only the activity method was used.

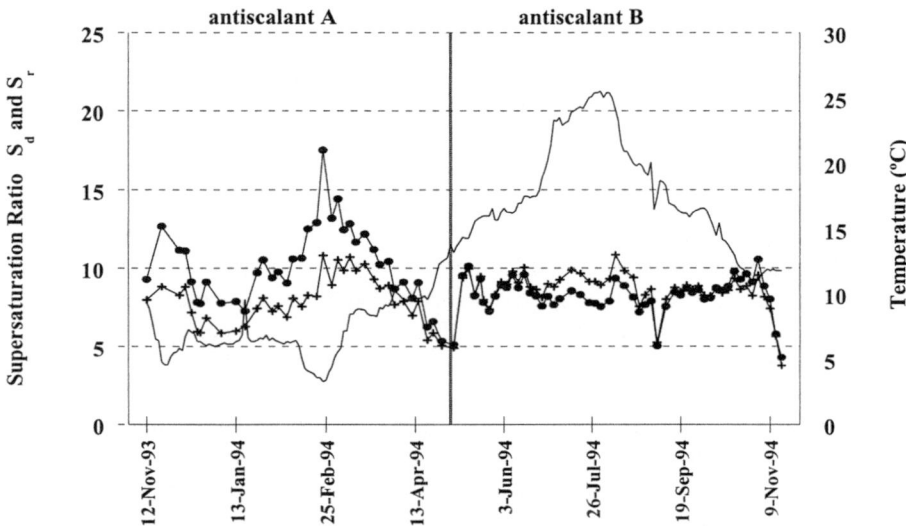

Figure 4.2: Comparison of supersaturation ratio calculated by the Du Pont (✚S_d) and activity methods (●S_r) and the RO feedwater temperature (——) during mode 1 (antiscalant A) and 2 (antiscalant B) for the RO pilot plant at 90% Recovery with H_2SO_4 acid.

Table 4.3: RO feedwater temperature, supersaturation ratio calculated by the Du Pont (S_d) and activity methods (S_r) and induction times calculated using S_r for operating modes (1 and 2) with antiscalant addition at the RO pilot plant.

Operating Mode	Temperature (°C)			Supersaturation Ratio S_d			Supersaturation Ratio S_r			Induction Time (min)				
	Min	Median	Max	Min	Median	Max	Min	Median	Max	Min	25%	Ave	Median	Max
1. Antiscalant A	3.3	6.5	12.8	5.0	7.9	10.8	5.4	9.7	17.5	2	9	36	18	410
2. Antiscalant B	11.0	16.6	25.5	3.8	8.7	10.9	4.3	8.2	10.5	10	15	42	18	555

4.4.1.2 Efficiency of Antiscalant A and B in Scale Prevention

Ideally, antiscalant addition should have prevented scaling. However, scaling was frequently found during the six months of operation with antiscalant A, up to 570mg/hr barium (refer Figure 4.3). Conversely, no scaling (<200mg/hr barium) was detected during the following six months of operation with antiscalant B.

Assuming the absence of antiscalant, induction times were calculated for the variation in supersaturation and temperature for the two operating modes and are presented in Figure 4.3. Induction times were found to range from 2-410 and between 10-555 minutes for mode 1 and 2, respectively. These short induction times were a consequence of the high supersaturation (median $S_r>8$) associated with this high recovery (90%) coupled to the added load of sulphate ions from the H_2SO_4 addition.

Although, conditions such as temperature, supersaturation and LSI under which the two antiscalants were applied, were not identical (refer Table 4.1 and 4.3) the median induction time for both modes was found to be equal at 18 minutes. However, the threat of scaling increases as the induction time decreases and was proportionally higher during operation with antiscalant A than with B when comparing the 25% quartile of the induction time data. Thus, in mode 1 the induction time of the bulk concentrate passing through the RO system in 25% of the cases was shorter than 9 minutes while in mode 2 it was less than 15 minutes (Table 4.3). Moreover, in November 1993 and March 1994 episodes occurred during operation with antiscalant A, where induction times were on the order of 2-20 minutes which coincided with scaling rates >200mg barium/hr (refer Figure 4.3). Whereas, no scaling was found during operation with antiscalant B despite 25% of induction times being lower than 15 minutes. Thus, antiscalant B was more successful in preventing scaling than A from a comparison of induction time data for the two modes.

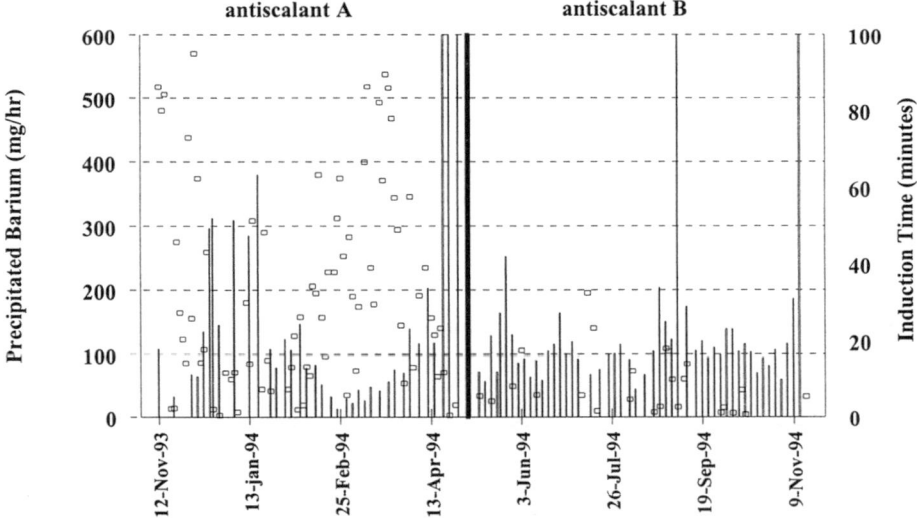

Figure 4.3:   Barium precipitated over the entire RO system calculated from the mass balance ❑ and the calculated induction time (❘) during mode 1 (antiscalant A) and 2 (antiscalant B) at 90% Recovery with H_2SO_4 acid.

Despite the presence of scale detected in the system during this time no irreversible decrease in the mass transfer coefficient was found or other adverse effects on RO operation [6]. However, the RO system was frequently cleaned to remove the barium sulphate scale. Although, cleaning may have successfully removed small crystal nuclei attached to the membrane and/or spacer, with such short induction times, it could not mitigate the precipitation of barium sulphate. The presence of an abundant number of nuclei could occur again after only a few minutes of operation following cleaning. These small nuclei could become attached to the membrane or spacer causing a snowball effect by continuing to grow in the highly supersaturated concentrate flowing through the system. Moreover, growth rates increase with supersaturation which would have exacerbated the scaling problem (refer Chapter 3 section 3.4.2.2).

The mechanism by which antiscalants prevent scaling is complex. Antiscalants are often reported to increase the induction time of a sparingly soluble salt e.g. He et al. [24,25]. However, this effect can be ascribed to preferential adsorption onto developing nuclei, prohibiting their outgrowth beyond the critical size required for growth [26]. Therefore, the more commonly accepted mechanism by which antiscalants prevent scaling is through crystal growth inhibition [1,26-30]. Thus, even if nucleation had occurred, antiscalant A should have prevented scaling.

Differences in antiscalant molecular weight and the functional groups present are known to have an important effect in the efficiency of antiscalant action. Both antiscalant A and B are polyacrylate polymers and higher molecular weight polyacrylates have been reported to be more effective [1]. This may explain in part why antiscalant B (mol. wt 3500) was more effective than the lower molecular weight antiscalant A (mol. wt 1600) in preventing scaling. Differences in scale inhibition may also have arisen due to the functional groups present, however, due to the proprietary nature of these chemicals, modifications of the attached functional groups are unknown.

The failure of antiscalant A may also have been as a result of the 1000:1 SO_4^{2-} to Ba^{2+} ion stoichiometry in the RO concentrate. The dominance of the SO_4^{2-} anion may repel an anionic inhibitor causing a time delay in the critical time for the inhibitor to attach to the crystal nuclei and inhibit its growth [1]. This effect could be reduced by optimisation of the antiscalant dose. Alternatively, although an antiscalant may inhibit the growth rate of a crystal it may promote nucleation [1,31].

Notwithstanding the failure of antiscalant A, the barium sulphate scale formed on the RO membranes was able to be removed during the frequent chemical cleanings [6]. A less cohesive scale may have been formed due to the dispersive property associated with antiscalants for particles coupled with weaker crystals being formed due to the incorporation of the antiscalant in the crystal lattice [1,32]. Alternatively, the crystals may have been only weakly attached to the membrane. More importantly the frequent cleaning prevented them from aging into a hard scale which would become resistant to conventional cleaning methods.

Although, antiscalant B was demonstrated to be effective in preventing scaling it was found to promote biofouling. This resulted in a very rapid increase in the feed pressure from 10 to the maximum admissible limit of 30 bar within 4-6 weeks in order to maintain productivity and recovery. Consequently, frequent cleaning of the RO system was also required in mode 2 [7,8].

4.4.2   Operation at 80% Recovery

Antiscalant addition was terminated to avoid biofouling, and operating mode 3 was adopted. Supersaturation (median $S_r = 3.9$) was more than halved in this mode by replacing sulphuric acid by hydrochloric acid and lowering the recovery to 80%. Operation under these conditions was found to be safe i.e. no scaling occurred for more than nineteen months of operation. Moreover, in the absence of antiscalant addition the cleaning of the RO system was delayed up to 8 months [6].

The corresponding induction times calculated for this mode are presented in Figure 4.4. In this mode the supersaturation was frequently below the limit of the relationship ($S_r = 3.16$) used for calculating induction time. Therefore, induction times were calculated at this limit for the corresponding RO feedwater temperature and represents the *minimum* induction time, indicated by arrows in Figure 4.4. Induction times were much longer in this mode and ranged from more than 276 hours to 49 minutes. This demonstrates that the induction time can vary dramatically for a given recovery, in this case by a factor greater than 340 times. No scaling was observed during this mode which suggests that induction times >82 hours (median for this mode) are long enough to prevent scaling.

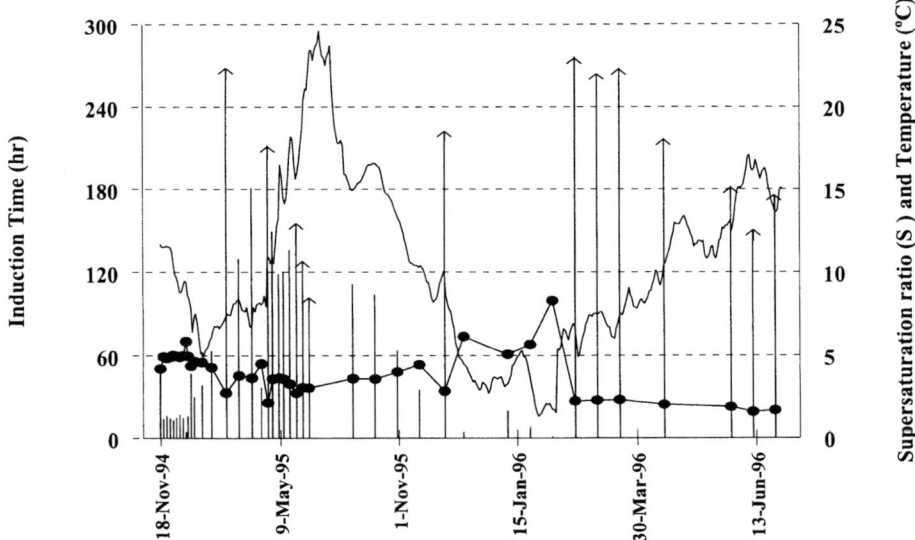

Figure 4.4:   Calculated induction time (❙) for supersaturation ratio (S_r ●) and RO feedwater temperature (——) during mode 3 at 80% recovery with HCl acid addition. Induction times indicated by (↑) represent the minimum induction time calculated for a S_r of 3.16 (limit for relationship used to calculate induction time) at the corresponding RO feedwater temperature.

4.4.3   Operation at 86-90% Recovery

In the following period (mode 4) the pilot plant was operated at increasing recovery steps, from 86-90% recovery to maximise the efficiency of the RO system and to determine the safe recovery limit, the point at which scaling occurred. BaSO$_4$ scaling was detected between 88-90% recovery via a barium mass balance (refer Figure 4.5) and an increase in pressure drop across the third stage [7]. The presence of barium sulphate scale was confirmed by membrane autopsy of one of the final elements of the last stage, by SEM and EDAX analysis.

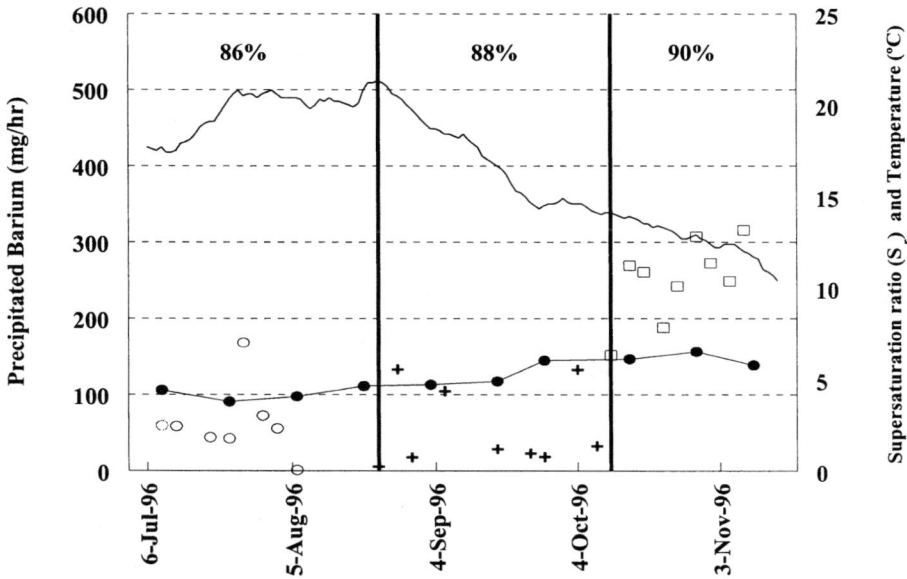

Figure 4.5: Mode 4; increasing recovery 86-90% with HCl acid addition. Calculated supersaturation ratio (S$_r$) ● for the feedwater temperature (—) and barium precipitated over the entire RO system calculated from the mass balance ○86% ✚88% ☐90% recovery.

The calculated induction times for mode 4 are presented in Figure 4.6. In general, an increase in recovery decreased the induction time. However, due to the variation in barium and sulphate concentration over time, the same induction time of 13 hours was found on one occasion at both 86 and 88% recovery. The shortest induction time at 88% recovery was circa 3 hours and after one week the recovery was increased to 90% recovery. Scale was detected from the barium mass balance one week later. It is not unlikely that scaling already started during operation at 88% recovery when the induction time was 3 hours. An increase in the differential pressure of the third stage was detected at 88% recovery. However, due to the sensitivity of the flow measurement system there will be a delay in the detection of scale in the RO system. Particularly as initially the scale will occur in a small localised region, typically in the final element of the last stage where supersaturation is highest due to the supersaturation gradient which exists over a membrane module.

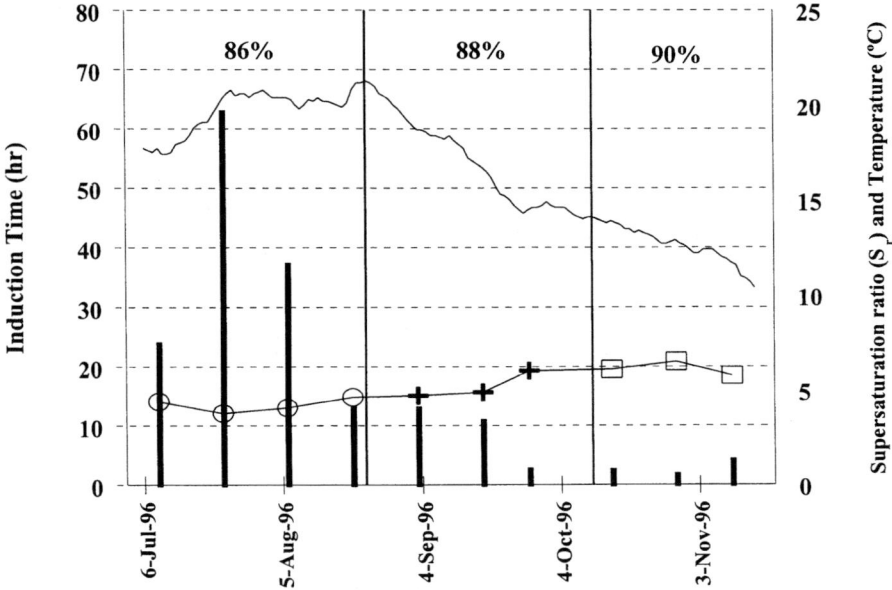

Figure 4.6: Decrease in the calculated induction time (**I**) corresponding to the supersaturation ratio (S_r) at ○86% ✚88% □90% recovery and to the feedwater temperature (——) with increasing recovery from 86-90% with HCl acid addition (mode 4).

4.4.4   Selection of Induction Time limits

Despite the limited data it appears that the critical induction time at which scaling occurs in the RO pilot plant for HCl addition is circa three hours. Whereas, when the induction time was greater than 10 hours, safe operation (no scale) was found for more than two months of operation. Based on these findings induction time limits were chosen to define safe and risky operation at the pilot plant. Whereby, an induction time of 10 hours or higher is assumed to be *safe* and will not result in scaling. To incorporate a safety margin an induction time of 5 hours was selected as the *risky* induction time limit rather than 3 hours. Operation below the risky limit could lead to scaling.

The induction time limits are based on batch test experiments and calculated from a function fitted for the RO concentrate and will most certainly have an inherent error due to the extrapolation of the line to higher supersaturation. However, more importantly batch tests can only simulate the behaviour of the bulk concentrate in the RO system and induction time are a function of feedwater characteristics only, such as pH, ionic strength etc. Real process conditions such as the presence of the membrane, membrane spacer, flow rate, and the detention time will greatly influence scaling in the RO system and cannot be simulated by such experiments. However, by correlating the calculated induction time to the occurrence of scaling in the RO system, these factors are taken into account.

An important element deciding scaling is if there is adequate contact time between the bulk RO concentrate and the membrane and/or spacer in the RO system for nucleation to occur. Results in this study suggest that an induction time of ≤5 hours for the bulk RO concentrate will probably result in scaling in the RO system. Although, the detention time of the bulk concentrate in the whole RO system is most likely on the order of only 1-1.5 minutes and in the last element not more than several seconds, a residence time distribution may exist and could result in a somewhat higher detention time for part of the flow in the system. However, of more consequence, the membrane and spacer "feel" a constant supersaturation and nucleation and/or the attachment of crystals has a longer time to proceed here than in the bulk solution. Moreover, nucleation most likely is faster in the RO system as it constitutes a much "dirtier" system than the batch test, with more nucleating surfaces available e.g. membrane and spacer which can catalyse heterogenous nucleation. The occurrence of deadlocks where the membrane and spacer touch are known to be sites of scaling [33]. These sites increase the supersaturation in localized regions and hence the driving force for nucleation. Thus, the equivalent induction time of 5 hours calculated for the bulk RO concentrate may be dramatically shorter in the RO system.

A safety margin of 2 hours was incorporated into the risky induction time limit to account for variations in RO process conditions e.g. flow rate with recovery and the synergistic effect of other foulants on the membrane/spacer which promote scaling. Changes in RO process conditions have an immediate effect on the scaling potential for example an increase in recovery lowers the linear flow velocities in the feed brine channel. Therefore, more time and opportunity is available for nucleation to occur and for crystals attachment to the membrane or spacer surface in the RO system. Moreover, there is less chance for the crystals to be removed from these surfaces by scouring. Whereas, the effect of other foulants on the scaling potential occurs after some period of RO operation and its effects are difficult to predict. For example the build up of a biofilm or fouling layer will lower flow velocities and can themselves act as sites for nucleation. Alternatively, the micro-organisms may cover crystals and prevent them from further growth.

4.4.5   Analysis of Safe Operating Modes

Subsequent to the pilot plant study of increasing recovery, the RO membranes were exchanged for ultra low pressure membranes to lower energy costs. The RO system was operated at 85% recovery (mode 5) for circa one year and no scaling was found. As expected the higher recovery of 85% resulted in a higher median supersaturation 4.7, in comparison to 3.9 found for the other safe operating mode 3 at 80% recovery (refer Table 4.4). Hence, the median induction time at 85% recovery of 14 hours for the bulk RO concentrate was at least 5.8 times shorter than that at 80% recovery of >82 hours.

However, the induction time can vary significantly for one recovery and hence the scaling potential, as observed in Figure 4.4 and 4.7 for 80 and 85% recovery, respectively. The induction time limits derived previously were used to examine the scaling potential of these two long term "safe" operating modes. It was assumed that the limits were valid for both types of RO membranes. The frequency of induction time for mode 3 and 5 with respect to the safe and risky induction time limits is presented in Table 4.5.

Table 4.4: RO feedwater temperature, calculated supersaturation ratio (S_r) and induction times for operating modes (3 and 5) without antiscalant addition at the RO pilot plant (* time in minutes).

Operating Mode	Temperature(°C)			Supersaturation Ratio S_r			Induction Time (hr)				
	Min	Median	Max	Min	Median	Max	Min	25%	Ave	Median	Max
3. HCl	1.3	9.4	24.6	1.6	3.9	8.3	49*	>16	>100	>82	>276
5. HCl	4.3	12.4	24.4	3.2	4.7	6.7	2	8	34	14	101

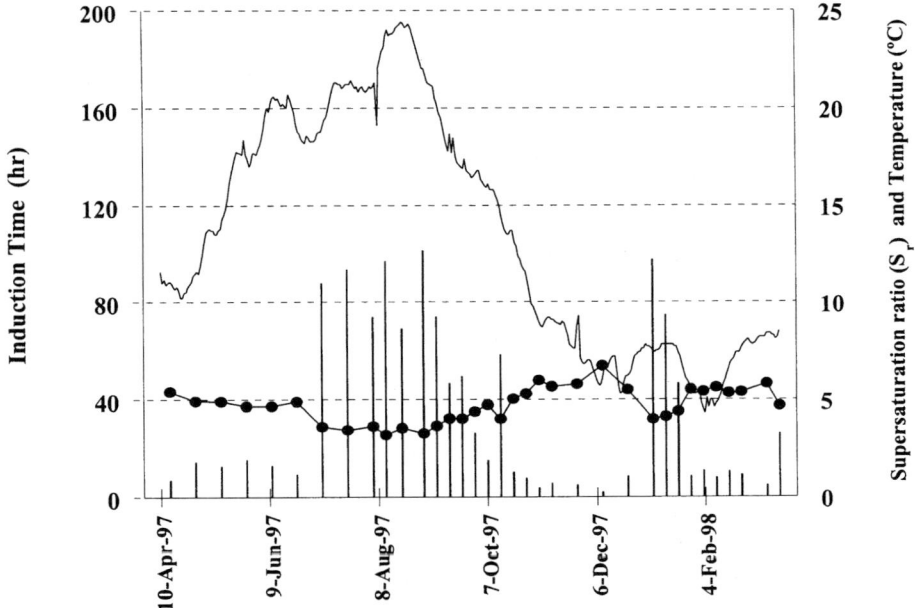

Figure 4.7: Calculated induction time (I) for supersaturation ratio (S_r ●) and RO feedwater temperature (——) during mode 5 at 85% Recovery with HCl acid addition.

Operation at the lower recovery of 80% resulted in 91% of the induction times being greater than the safe induction time limit of 10 hours. Nevertheless, 7% of the induction times (three occasions) were shorter than the risky induction time limit of 5 hours and on one occasion the potential for scale was serious as the induction time was only 49 minutes. Although, no scale was detected, crystals may have formed in the bulk concentrate flowing through the system but which did not attach to the membrane or spacer. Therefore, they were simply flushed out of the system along with the bulk concentrate.

Furthermore, inspection of Figure 4.4 shows these three occasions occurred in isolation and shortly after the induction time increased to well above the safe induction time limit e.g. in the case of the shortest induction time of 49 minutes to >276 hours. Moreover, this low induction time was a consequence of the very low RO feedwater temperature of 1.3°C (which increased the supersaturation). It has been reported that at very low temperatures the water may be too viscous to nucleate [9]. The viscosity is almost doubled for the temperature drop from 25 to 1.3°C and this may have extended the induction time preventing scaling.

Operation at the higher recovery of 85%, shifted the induction time distribution to a lower frequency of induction time (66%) above the safe induction time limit. The scaling risk increased as expected with a recovery increase, as 11% of the induction times were found to be below the risky induction time limit. As in the case of mode 3 the four occasions for which the scaling potential was serious occurred in isolation. No scaling was detected at these times, most likely due to the same reasons discussed previously.

Table 4.5: Comparison of induction time data during the safe operating modes 3 and 5 at 80 and 85% recovery, respectively, in the following induction time zones; safe operation i.e. no scaling risk at >10 hours, low scaling risk 5-10 hours and high scaling risk at<5 hours.

Frequency Induction Time (hrs)	Recovery	
	80% (mode 3)	85% (mode 5)
<5	7	11
5-10	2	23
>10	91	66

4.4.6.   Safe Supersaturation Limits

In practice it is easier to monitor the scaling potential over time for the RO system in terms of safe and risky supersaturation limits (calculated by the activity method) rather than the previously derived induction time limits. However, due to the double effect of temperature on induction time *and* supersaturation, safe and risky supersaturation limits will vary with temperature. Therefore, the selected induction time limits of 10 and 5 hours determined for this RO pilot plant were translated into (i) *safe supersaturation limits* and (ii) *risky supersaturation limits* for the 5-25°C temperature range (refer Table 4.6), respectively. This temperature range corresponds to the average seasonal temperature fluctuations experienced at the pilot plant.

At the lower RO feedwater temperature of 5°C the resultant safe supersaturation (SS_r) limit was 5.4 and as the scaling potential increases with a temperature increase the SS_r decreased to 4.6 for a feedwater temperature increase to 25°C. Operating below or equal to the S_rS presents no scaling risk in the RO system. Whereas, the risky supersaturation limit (RS_R) where scaling most likely will occur is 6.0-5.0 for the same temperature range. Operation between the safe and risky limits most likely will not result in scaling in the RO system and represents a low scaling risk. The proposed supersaturation limits can be used to assess the risk of scale for a proposed operating mode. In addition, the supersaturation ratio calculated on a regular basis can be used to monitor the scaling potential over time. This is especially useful as changes in water quality and temperature cause a wide variation in the actual supersaturation ratio achieved over time at a fixed recovery. When the supersaturation is found to exceed the risky scaling limit (RS_r) for the corresponding temperature, the system recovery could be temporarily lowered to prevent scaling.

The general validity of the proposed limits, however, should be confirmed with pilot or full scale studies with different feedwater compositions, RO elements, recoveries.

Table 4.6:   RO operation below or equal to the safe supersaturation limits (SS_r) will result in no scaling risk, operation above or equal to the risky supersaturation limits (RS_r) represents a high scaling risk while operation between the limits represents a low scaling risk for temperatures of 5 - 25 °C.

Supersaturation Ratio (S_r) Limit	Temperature (°C)				
	5	10	15	20	25
Risky supersaturation limit *i.e.* high scaling risk at $S_r >$ RS_r (Induction time = 5hr)	6.0	5.7	5.5	5.2	5.0
Low scaling risk at $SS_r > S_r < RS_r$	5.4-6.0	5.2-5.7	5.0-5.5	4.8-5.2	4.6-5.0
Safe Supersaturation Limit *i.e.* no scaling risk at $S_r < SS_r$ (Induction time = 10hr)	5.4	5.2	5.0	4.8	4.6

4.5   Conclusions

The temperature limitation of 25 °C coupled with solubility being underestimated at 25 °C by the Du Pont method leads to an error in quantifying supersaturation in the RO concentrate by -40 to +15% for the RO feedwater temperature range of 3-25 °C, respectively. Therefore, the more accurate activity method was employed to quantify supersaturation which corrects for variations in the RO feedwater temperature.

The median induction times calculated for the 90% RO concentrate with antiscalant A and B addition were very short at 18 minutes for both modes. Scaling episodes during operation with antiscalant A coincided with shorter induction times, 2-15 minutes. The failure of this antiscalant suggests that induction times in this range most likely will result in scaling. Therefore, operating within this range requires the addition of a more effective antiscalant. Although, antiscalant B was effective it caused serious biofouling.

Operation at 80% recovery, with HCl addition and no antiscalant reduced the median supersaturation by more than half. This extended the median induction time to >82hours, more than 270 times longer than during mode 1 and 2. No scaling was evident despite the high supersaturation.

Safe and unsafe induction time limits, were determined by correlating induction time to periods when scaling did and did not occur in the RO system during operation mode 4. In this mode the recovery was increased step wise from 86, 88 to 90% recovery. Induction times shorter than 5 hours most likely result in scaling and induction times greater than 10 hours will be safe. Safe long term operation at 80 and 85% recovery can be attributed to operation with 90% and 66% of the induction times above the safe induction time limit, respectively.

Safe and risky supersaturation limits for the pilot plant were derived corresponding to induction times of 10 and 5 hours, respectively. In the temperature range of 5-25°C, these limits range from 5.4-4.6 (safe) to 6.0-5.0 (risky). Use of these limits gives a more dynamic approach to avoiding scaling while optimising the RO system recovery. This involves monitoring the supersaturation and temperature over time which allows the scaling potential to be assessed and recovery adjustment to maintain operation within these limits.

However, the validity of the suggested induction times and corresponding supersaturation limits needs to be confirmed with pilot studies using different reverse osmosis elements and with feedwater of different quality at various recoveries before they can be considered for general use

Symbols

A	pre-exponential factor (nuclei/cm^{3}s)
A_3	slope in Equation 4.14 for temperature correction factor
A_1	intercept in Equation 4.14 for temperature correction factor/constant in Equation 4.9 (1/s)
$a_{i(aq)}$	activity of ion i (mol/L)
B	defined by Equation 4.8 (-)
BA	mass of barium precipitated in RO system (mg/hr)
C	concentration of barium in feedwater $_{(f)}$, brine $_{(b)}$, and permeate $_{(p)}$ (mg/m^3)
CF	concentration factor (-)
E	constant in Equation 4.7
E_a	activation energy (J/mol)
F	geometrical shape factor (for spherical nucleus $16\pi/3$)
G	crystal growth rate (m/s)
ΔG_{cr}	free energy change for the critical cluster size (J/molecule)
J_n	nucleation rate (nuclei/cm^{3}s)
k	Boltzmann constant (J/K molecule)

K_C	$BaSO_4$ solubility product $(mol/L)^2$
K_{sp}	$BaSO_4$ thermodynamic solubility product $(mol/L)^2$
M	Molal (mol/L water)
N_a	Avogadro's number (molecules per mol)
Q	flow of RO feedwater $_{(f)}$, brine $_{(b)}$, and permeate $_{(p)}$ (m^3/hr)
R	universal gas constant (J/K mol)
S_r	supersaturation ratio using activities (-)
S_d	supersaturation ratio for Du Pont(-)
t_{ind}	induction time (s)
T	absolute temperature (K)
$T_{°C}$	temperature in degrees celcius (Equation 4.3)
V	volume of solution (m^3)
V_m	molar volume of solid (cm^3/mol)
Y	recovery (-)
[]	concentration (mol/L)

Greek Symbols

α_1	first detectable crystallised volume fraction (V_{macro}/V) (-)
γ_+	ionic activity coefficient of cation (-)
γ_-	ionic activity coefficient of anion (-)
$f(\theta)$	contact angle factor (-)
σ	interfacial tension (J/m^2)

Abbreviations

EDAX	Energy Dispersion by X-Ray
LSI	Langelier Saturation Index
SEM	Scanning Electron Microscope

References

1.   M.C. Van der Leeden, The role of polyelectrolytes in barium sulfate precipitation, Ph.D Thesis, TU Delft, 1991.

2.   Permasep Engineering Manual (PEM), Du Pont de Nemours & Co. 1982.

3.   C.C. Templeton, Solubility of barium sulphate in sodium chloride from 25° to 95°C. J. of Chem. and Eng.Data, 5 (1960) 514-516.

4.   J.A.M. Paassen, J.C.Kruithof, S.M.Bakker and F. Kegel-Schoonenberg, Integrated multi-objective membrane systems for surface water treatment: pre-treatment of nanofiltration by riverbank filtration and conventional ground water treatment, Desal., 118 (1998), 239-248.

5.   J.P. Van der Hoek, P.A.C.Bonné and E.A.M. Van Soest, Fouling of reverse osmosis membranes: The effect of pretreatment and operating conditions. Proceedings of the American Waterworks Association 1997 Membrane Technology Conference, February 23-26, New Orleans, U.S.A. 1029-1041.

6.   J.P. Van der Hoek, P.A.C. Bonné and E.A.M. Van Soest and A. Graveland, Application of hyperfiltration at the Amsterdam Waterworks: Effect of pretreatment on operation and performance, Proceedings of the American Waterworks Association 1995 Membrane Technology Conference, August 13-16 Reno, U.S.A. 277-294.

7.   J.P. Van der Hoek, P.A.C.B onné and E.A.M. Van Soest, Reverse Osmosis: Finding the balance between fouling and scaling, Proceedings of the 21st IWSA World Congress 1997, September 20-26, Madrid, Spain.

8.   J.P. Van der Hoek, J.A.M. H.Hofman, P.A.C. Bonné, M.M. Nederlof and H.S. Vrouwenvelder, RO treatment: Selection of a pre-treatment scheme based on fouling characteristics, and operating conditions based on environmental impact, Proceedings of the American Waterworks Association 1999 Membrane Technology Conference, February 28-March 3 Long Beach,U.S.A.

9.   A.S. Myerson, Handbook of Industrial Crystallization. Butterworth-Heinemann, 1993.

10.   O. Söhnel and J. Garside, Precipitation - Basic principles and industrial applications, Butterworth-Heinemann, 1992.

11.   J.W.Mullin, Crystallization, 3rd Edition, Butterworth-Heinemann, 1993.

12.   O. Söhnel and J.W. Mullin, Interpretation of crystallization induction periods, J of Colloid and Interface Sci., 123 (1988) 43 - 50.

13.   A.E. Nielsen, Kinetics of precipitation Pergamon Press, New York, 1967.

14.   S.F.E. Boerlage, M.D. Kennedy, G.J. Witkamp, I. Bremere, J.P. Van der Hoek, and J.C. Schippers, Stable barium sulphate supersaturation in reverse osmosis, J. Mem. Sci., 179, (2000) 53-68.

15.   S.F.E. Boerlage, M.D. Kennedy, G.J. Witkamp, J.P. Van der Hoek and J.C. Schippers $BaSO_4$ Solubility in reverse osmosis concentrates, J. Mem. Sci., 159 (1999) 47-59.

16   W.R. Linke, and A. Seidell, Solubilities - Inorganic and metal-organic compound Vol II American Chemical Society, Washington DC, 1965.

17.   J.W. Zhang and G.H. Nancollas, Mechanisms of growth and dissolution of sparingly soluble salts. in: Reviews in Mineralogy, 23 (1990) 365 - 396.

18.   S. He, J.E.Oddo, and M.B.Tomson, The nucleation kinetics of barium sulfate in NaCl solutions up to 6M and 90°C, J. of Colloid and Interface Sci, 174 (1995) 319-326.

19.   M.C. van der Leeden, D. Verdoes, D. Kashchiev and G.M. van Rosmalen, Induction time in seeded and unseeded precipitation, in J. Garside, R.J. Davey and A.G. Jones [eds], Advances in Industrial Crystallization, Butterworth-Heinemann, Oxford (1991) 31-46.

20.   D. Kashchiev, Nucleation basic theory with applications, Butterworth Heinemann, ISBN 0 7506 4682 9, 2000.

21.   D. J. Gunn, and M.S.Murthy, Kinetics and mechanisms of precipitations, Chem. Eng. Sci. 27 (1972) 1293-1313.

22.   J.A.M.H. Hofman, AWS, Personal Communication.

23.   F.Van der Ham, Pitzer Model from Laboratory for Process Equipment, Delft University of Technology, Leeghwaterstraat 44, 2628 CA Delft, The Netherlands.

24.   S. He, J.E. Oddo, and M.B. Tomson, The inhibition of gypsum and barite nucleation in NaCl brines at temperatures from 25 to 90°C. App. Geochem., 9, (1994) 561-567.

25.   S. He, J.E.Oddo and M.B.Tomson, The nucleation kinetics of strontium sulfate in NaCl solutions up to 6M and 90°C with or without inhibitors, J. of Coll. and Interf. Sci, 174 (1995) 327-335.

26.   M.P.C.Weijen and G.M.Van Rosmalen, The influence of various polyelectrolytes on the precipitation process, Desal. 54 (1985), 239-261.

27.   S.T. Liu and G.H. Nancollas, A kinetic and morphological study of the seeded growth of calcium sulphate dihydrate in the presence of additives, J. of Coll. and Interface Sci., 52 (1975) 593-601.

28.   S.T. Liu and G.H. Nancollas, The crystal growth and dissolution of barium sulphate in the presence of additives, J. of Colloid and Interface Sci., 52 (1975) 582-592.

29.   M.C. van der Leeden, D. Kashchiev and G.M. van Rosmalen, Precipitation of barium sulfate: Induction time and the effect of an additive on nucleation and growth. J. of Colloid and Interface Sci, 154, (1992) 338-350.

30.   M.C. van der Leeden and G.M. van Rosmalen, Aspects of additives in precipitation processes: performance of polycarboxylates in gypsum growth prevention. Desal., 66, (1987) 185-200.

31.   N. Eidelman, R. Azoury and S Sarig, Reversal of Trends in Impurity Effects of Crystallization Parameters, J. of Crystal Growth, 74 (1986) 1-9.

32.   Z. Amjad, RO Antiscalants: Advances in scaling and deposit control for RO systems, ultrapure water, 4 (1987) 34-38.

33.   A. G. Pervov, Scale formation prognosis and cleaning procedure schedules in RO system operation, Desalination, 83 (1991) 77-118.

5

Development of the Modified Fouling Index using Ultrafiltration Membranes (MFI-UF)

Chapter 5 is based on: "The Modified Fouling Index using Ultrafiltration Membranes (MFI-UF): Characterisation, Filtration Mechanisms and Proposed Reference Membrane" by S.F.E. Boerlage, M.D. Kennedy, M. R. Dickson and J.C. Schippers. Submitted to *Journal of Membrane Science*, (2001).

Contents

Abstract

The existing Modified Fouling Index ($MFI_{0.45}$), based on cake filtration, uses a 0.45μm microfilter to measure the particulate fouling potential of a feedwater. However, it is not sensitive to small colloidal particles. To incorporate these particles the MFI using ultrafiltration (UF) membranes was proposed. In this study a suitable reference membrane for the MFI-UF test was investigated using polysulphone (PS) and polyacrylonitrile (PAN) UF membranes of 1-100 kDa molecular-weight-cut-off in tap water experiments. The stability of the MFI-UF value over time for the PAN and PS membranes and the influence of MWCO on the MFI-UF obtained were examined. Field emission scanning electron microscopy (FESEM) of the membrane surfaces and an analysis using the constant pressure blocking and cake filtration models were carried out. A stable MFI-UF* was found for PAN membranes. Whereas, the MFI-UF continuously decreased over time for PS membranes. The measured MFI-UF* (2000-13 300 s/l^2) were significantly higher than the $MFI_{0.45}$ expected for tap water, (1-5 s/l^2), indicating smaller particles were retained. FESEM showed PAN membranes were homogeneously permeable, resulting in a stable MFI-UF* value as cake filtration was proven to be the dominant mechanism. FESEM showed the 1,2, and 5 kDa PS membranes were heterogeneously permeable and sparsely porous resulting in (i) artificially high MFI-UF*and (ii) limited cake filtration. Hence the MFI-UF never stabilised. The MFI-UF* appeared MWCO independent within the 3-100 kDa MWCO range as most likely the cake itself acts as a second membrane, determining the size of particles retained and the resultant MFI-UF*.

5.1   Introduction

In pressure driven membrane filtration processes; reverse osmosis, nano, ultra and micro-filtration, the maintenance of a high permeate flux with the lowest possible energy requirement is the principal design and operation goal. However, in the application of these processes for potable water production, especially from surface water sources, a decline in permeate flux is often found due to particulate fouling. Particulate fouling refers to the deposition of colloids, suspended solids, and microbial cells, present in the feedwater, onto the membrane or spacer in the feed-concentrate channel. The process by which particulate fouling leads to flux decline in microfiltration [1] and ultrafiltration [2,3] is often described by blocking and cake filtration mechanisms. Deposition of colloids on reverse osmosis and nanofiltration (NF) membranes is expected to be mainly by the latter mechanism. Blocking generally refers to particles completely sealing a pore ("complete blocking") or entering directly into a pore ("standard blocking") resulting in the loss of filtration area. Whereas, cake filtration refers exclusively to the deposition of particles onto the membrane surface forming a cake.

To maintain membrane productivity, an increase in the applied pressure in combination with membrane cleaning is required, both of which increase energy costs. Therefore, methods to measure the particulate content of a feedwater and to predict membrane fouling are important tools in the control of particulate fouling, both at the design stage and for monitoring during plant operation. Presently, the most widely applied fouling indices to measure the particulate fouling potential of membrane filtration feedwater are the Silt Density Index (SDI), and the Modified Fouling Index ($MFI_{0.45}$). In both tests, the water is filtered through a 0.45µm microfiltration membrane in dead-end flow at a constant pressure. Since, the SDI is not based on a distinction between blocking and cake filtration mechanisms occurring during the test, it can not be used as part of a mathematical model to predict the rate of flux decline due to particulate fouling.

Conversely, the MFI is based only on the cake filtration mechanism and is dependent on particle size through the Carmen-Kozeny equation for specific cake resistance [4,5]. Thus, in general, smaller particles present in the cake result in higher MFI values. Assuming cake filtration is the dominant mechanism in particulate fouling in RO and NF, the MFI can be used as a basis for modelling flux decline or pressure increase to maintain constant capacity in membrane systems [5]. Calculations based on the cake filtration mechanism show that measured $MFI_{0.45}$ for RO feedwater were far too low to explain the flux decline rates observed in practice [5]. It was therefore hypothesised that smaller colloidal particles were responsible for the observed flux decline rates [5].

To more accurately measure and predict particulate fouling, the MFI is being developed using ultrafiltration (UF) membranes to incorporate fouling due to these smaller colloidal particles (referred to as MFI-UF). This research focuses on proposing a reference UF membrane for the MFI-UF test. Principal factors which need to be considered in proposing a reference membrane are membrane material, pore size, surface porosity and the occurrence of cake filtration. Furthermore, ultrafiltration membranes have lower fluxes than the existing MFI membrane, therefore, the time taken for cake filtration to occur, or the MFI-UF value to stabilise over time, may be prohibitively long. Whereas, for practical use, it is desirable to carry out the MFI-UF measurement in as short time as possible. A final consideration is that while the existing MFI test uses inexpensive disposable microfilters UF membranes are expensive. Therefore, for the test to be viable in practice, a simple method is required to clean the UF membranes to allow membrane reuse.

Various materials are used in manufacturing UF membranes e.g. cellulose blends, polyamide, polysulphone etc. Membrane material charge, hydrophobicity and resultant structure arising from the manufacturing process may play a role in the time required for cake filtration to occur. In addition, the efficiency of membrane cleaning will depend on the strength of electrostatic charge and hydrophobic interactions between the foulant and the membrane material.

Ideally, UF membranes with smaller pore sizes are desirable in the MFI-UF test. Ultrafiltration membrane manufacturers frequently characterise their membranes using the "cut-off" concept rather than pore size. The nominal molecular-weight-cut-off (MWCO) is a performance related parameter, defined as the lower limit of a solute molecular weight e.g. dextran for which the rejection is 95-98% [6]. As the MWCO decreases the mean pore diameter for most ultrafiltration membranes has been found to decrease [7]. Therefore, UF membranes of a smaller MWCO in the MFI-UF test should capture smaller colloidal particles from a feedwater. However, the MWCO may be sharp or diffuse *i.e.* there is a range of MWCO and in reality MWCO is only a rough indication of the membranes ability to remove a given compound as molecular shape, polarity and interaction with the membrane affect rejection [8,9]. Moreover, membrane surface characteristics e.g. surface porosity and pore size distribution may influence the apparent size of particles retained. Therefore, the MWCO rating may not reflect the particle size retained.

The extent and occurrence of blocking and cake filtration mechanisms may also be influenced by membrane surface characteristics. Surface porosity, in particular is important as UF membranes with a low surface porosity have been found to be irregularly permeable and susceptible to pore blocking [10]. As a result, cake filtration (the basis of the MFI), may take a long time to develop with such membranes. However, information on membrane surface morphology e.g. surface porosity and pore size distribution is not readily available from membrane manufacturers. Therefore, UF membranes need to be characterised in order to determine the influence of surface morphology on the MFI-UF value obtained and on the filtration mechanisms. One of the most successful techniques for surface characterisation of ultrafiltration membranes is field emission scanning electron microscopy (FESEM) [7,11-13].

This research investigates the application of hollow fibre ultrafiltration membranes in the MFI-UF test. A broad MWCO range (1 to 100 kDa) of commercially available membranes manufactured from two different membrane materials will be examined. The influence of (i) membrane material on the stability of the MFI-UF value over time and on the efficiency of cleaning and (ii) the effect of MWCO on the MFI-UF value will be investigated. FESEM will be employed to determine surface morphology characteristics of the membranes e.g. surface porosity. Surface morphology characteristics will be used to explain the occurrence of blocking and cake filtration mechanisms, differences in the behaviour of the MFI-UF value over time and the final (real) MFI-UF value obtained for a membrane. Finally, a reference UF membrane will be proposed for application in the MFI-UF test.

5.2   Background

In this section blocking and cake filtration mechanisms, which are principally responsible for the retention of particles during membrane filtration, are reviewed from a historical perspective.

5.2.1   Cake Filtration

Cake filtration is based on the fundamental equation for the rate of flow through a porous medium:

$$J = \frac{dV}{A\,dt} = K\,\frac{\Delta P}{\eta\,L}$$ ...(5.1)

where J is the linear fluid velocity (of volume V flowing in time t through cross section area A) across the porous medium of length (L) for the pressure gradient (ΔP) and the viscosity of the fluid (η). K refers to the permeability of the porous medium. Equation 5.1 is a combination of the Laws of Darcy (1856) and Poiseuille (1840-42) for the rate of flow through a sand bed and through circular capillary tubes, respectively. Both laws assume laminar flow. The inverse of K/L, termed specific permeability, is used to define the resistance (R = L/K) to flow. In early cake filtration studies, Sperry (1916 [14]) used resistance analogously to Ohms Law, dividing resistance to filtration into; an initial resistance due to the filter media (R_m) and that due to the cake deposited onto the filter (R_c). Thus, Equation 5.1 can be rewritten as:

$$\frac{dV}{A\,dt} = \frac{\Delta P}{\eta\,(R_m + R_c)}$$ ...(5.2)

Permeability of the clean filter media (R_m) is a function of filter properties such as filter thickness (Δx), surface porosity (ε), pore radius (r_p), and tortuosity (τ) and can be defined using Poiseuille's Law:

$$R_m = \frac{8\,\Delta x\,\tau}{\varepsilon\,r_p^2}$$ ...(5.3)

The resistance in series model was employed by many researchers including Ruth [15] and Carmen [16,17] in developing cake filtration theory. The cake resistance (R_c) component in (membrane) filtration can be defined following the Ruth equation [15], using the concept of "specific cake resistance" per unit weight (α) (Equation 5.4). Ruth showed that the resistance of the cake formed during constant pressure filtration is proportional to the amount of cake deposited at the filter medium provided the retention of particles and α are constant:

$$R_c = \frac{V}{A} \times \alpha\,C_b$$ ...(5.4)

where C_b is the concentration of particles per unit volume of filtrate. The specific cake resistance is constant for incompressible cakes under constant pressure filtration and can be calculated according to the Carman-Kozeny relationship (Equation 5.5) [16,17]. Carmen [16, 17] derived Equation 5.5 for the specific resistance of a cake composed of spherical particles of diameter d_p from the Kozeny equation including a factor for tortuosity of the voids within the cake. According to the Carmen relationship a reduction in the porosity of the cake (ε) or a decrease in particle diameter size (d_p) increases the specific resistance of the deposited cake.

$$\alpha = \frac{180\,(1-\varepsilon)}{\rho_p\,d_p^{\,2}\,\varepsilon^3} \qquad\qquad\qquad ...(5.5)$$

Combining Equations 5.2 and 5.4 and integrating at constant ΔP from $t = 0$ to $t = t$, assuming time independent permeability and uniform porosity characteristics throughout the depth of the cake (i.e. no compression of the cake), results in the well known filtration equation:

$$\frac{t}{V} = \frac{\eta\,R_m}{\Delta P\,A} + \frac{\eta\,\alpha\,C_b}{2\Delta P\,A^2}\,V \qquad\qquad ...(5.6)$$

Equation 5.6 gives a straight line when t/V is plotted against V and has been widely applied since suggested by Underwood in 1926 [18] to test for cake filtration and to obtain information on the permeability of the cake deposited. Carmen defined the gradient of the line (b) as [17]:

$$b = \frac{\eta\,\alpha\,C_b}{2\Delta P\,A^2} \qquad\qquad\qquad ...(5.7)$$

The gradient of the line was adopted by Schippers [5] to define the Modified Fouling Index (MFI) as an index of the fouling potential of a feedwater containing particles, when fixed reference values are used for ΔP (2 bar), η ($\eta_{20°C}$) and A (13.8×10^{-4} m^2). In the MFI (Equation 5.8), the product of the specific resistance of the cake and the concentration of particles in the feedwater is taken to equal I the fouling index, and is assumed to be independent of pressure. An advantage of using I is that in most cases it is impossible to determine C_b and α accurately. The fouling index I is a function of the dimension and nature of the particles (through Equation 5.5) present in a feedwater and directly correlated to their concentration [4].

$$\text{MFI} = \frac{\eta\,\alpha\,C_b}{2\Delta P\,A^2} = \frac{\eta\,I}{2\Delta P\,A^2} \qquad\qquad ...(5.8)$$

5.2.2   Blocking Filtration

In 1936 Hermans and Bredée considered other filtration mechanisms that could retain particles during constant pressure dead end filtration and introduced the blocking laws. In their study, the filter medium was modelled as parallel Poiseuille capillary tubes [19]. The complete blocking law physically described the complete sealing of a tube with no cake formed on the filter. Whereas, the standard blocking law described the narrowing of a tube by the internal deposition of particles on the tube wall which causes a progressive restriction of the free tube volume. Hermans and Bredée identified a third blocking law empirically, which they termed "intermediate" between complete and standard blocking. However, they failed to present a physical model to visualise the process. Despite this, they characterised cake filtration and the three blocking mechanisms as conforming to the basic equation:

$$\frac{d^2t}{dV^2} = k\left(\frac{dt}{dV}\right)^n \qquad\qquad ...(5.9)$$

where k and n are filtration constants. Values of n of 0, 1, 1.5 and 2 define cake, intermediate, standard and complete blocking filtration, respectively. Hermans and Bredée proposed various plots of filtration data to test which filtration law was obeyed. Linearity of the data in the dV/dt vs V, t/V vs t and dt/dV vs V plots was proof of the respective laws; complete, standard and intermediate blocking [19].

Gonsalves [20] questioned the physical models used by Hermans and Bredée to derive these laws. In particular, Gonsalves criticised the indiscriminate application of the standard blocking law to analyse initial filtration data which was commonplace at that time. However, his experiments confirmed that standard blocking was followed for a significant part of the remaining filtration run prior to cake filtration. Furthermore, he gave a full description of the four filtration laws for (i) volume as a function of time and (ii) flow rate (Q) as a function of (a) time and (b) volume.

Hermia [21,22] re-analysed the filtration laws of Hermans and Bredée and developed a physical model to describe the transition mechanism from pore blocking to cake formation i.e. the intermediate blocking law. This latter law was expressed as the probability of a particle completely blocking a pore with particles allowed to settle on other particles deposited previously i.e. cake filtration. Although, he presented a full description of the four filtration laws in their integrated form as Gonsalves, he recommended [21] the same plots to confirm which filtration mechanism was followed as Hermans and Bredée. A summary of the integrated Equation 5.9 for the filtration laws according to Hermia are given in Table 5.1 for V = f (t) and Q = f (t).

In classic cake filtration theory the resistance of the membrane is considered constant. However, considering that blocking filtration may occur prior to cake filtration, Equation 5.2 may be modified to include R_b to include the resistance of the blocked filter. Heertjes recognised that R_m does not have to be constant [23] and if R_b reaches a stable value prior to cake filtration then, Equation 5.2 may still be valid.

Table 5.1: Equations for constant pressure Blocking and Cake filtration Laws (** k' for standard blocking = k √dv/dt) [21]

Function	Complete Blocking	Standard Blocking**	Intermediate	Cake Filtration
$\dfrac{d^2t}{dV^2} = k\left(\dfrac{dt}{dV}\right)^n$	n=2	n=3/2 $$\dfrac{d^2t}{dV^2} = k'\left(\dfrac{dt}{dV}\right)^n$$	n=1	n=0
$V = f(t)$	$kV = \dfrac{dV_o}{dt_o}\left(1 - \exp^{-kt}\right)$	$\dfrac{k}{2}t = \dfrac{t}{V} - \dfrac{dt_o}{dV_o}$	$kV = \ln\left(1 + k\dfrac{dV_o}{dt_o}\,t\right)$	$\dfrac{kV}{2} = \dfrac{t}{V} - \dfrac{dt_o}{dV_o}$
$dV/dt = f(t)$	$\dfrac{dV}{dt} = \dfrac{dV_o}{dt_o}\exp^{-kt}$	$\dfrac{dV}{dt} = \dfrac{\dfrac{dV_o}{dt_o}}{\left(1 + \dfrac{k}{2}\dfrac{dV_o}{dt_o}t\right)^2}$	$kt = \dfrac{dt}{dV} - \dfrac{dt_o}{dV_o}$	$\dfrac{dV}{dt} = \dfrac{\dfrac{dV_o}{dt_o}}{\sqrt{1 + 2k\left(\dfrac{dV_o}{dt_o}\right)^2 t}}$

5.3   Materials and Methods

5.3.1   Membranes

Eleven commercially available hollow fibre ultrafiltration membranes with MWCO ranging from 1 to 100 kDa were investigated. (The hollow fibre configuration was used in this study as other membrane configurations such as flat sheet were not available for such a wide MWCO range). Two of the membrane series were manufactured from polysulphone and one series from polyacrylonitrile, abbreviated as PS A, PS B and PAN, respectively. The PS B and PAN membrane series were obtained from the same manufacturer. Membrane specifications and the nominal MWCO, as rated by the manufacturer, are summarised in Table 5.2. All membranes tested were new or their clean water flux was restored to ≈100% prior to testing. The clean water flux (CWF) was measured in the MFI-UF equipment (refer Figure 5.1) using RO permeate and calculated according to Equation 5.10 corrected to 20°C and a transmembrane pressure of 1 bar.

$$CWF = \frac{\eta_T}{\eta_{20°C}} \frac{Q}{A \, \Delta P} \qquad\qquad ...(5.10)$$

where Q is the clean water flow at temperature T, and A is the membrane surface area.

Table 5.2: Specifications of the ultrafiltration membranes

Membrane Material	MWCO kDa	Membrane Area (m²)	Clean Water Flux (l/m²/hr) at 20°C and 1 bar	Fibre Length (m)	Inner Fibre Diameter (mm)
Polysulphone manufacturer A	1	0.46	67	0.60	1.1
	2	0.46	173	0.60	1.1
	5	0.46	95	0.60	1.1
	10	0.46	170	0.60	1.1
	50	0.46	304	0.60	1.1
	100	0.09	250	0.45	1.1
Polysulphone manufacturer B	3	0.20	95	0.25	0.8
	6	0.20	200	0.25	0.8
Polyacrylonitrile manufacturer B	6	0.20	55	0.25	0.8
	13	0.20	170	0.25	0.8
	50	0.20	400	0.25	0.8

5.3.2   Determination of MFI-UF

Delft tap water pretreated by conventional treatment processes i.e. coagulation, sedimentation and filtration was used as the feedwater. The MFI-UF value was determined with the UF membrane connected in dead end flow (retentate outlet closed) in the MFI-UF equipment (Figure 5.1). The feedwater was pumped on-line to the UF membrane inlet in lumen side filtration at a constant transmembrane pressure (1.0 bar), using a pressure reducing valve. Flow was measured by a micro-oval flowmeter (Flowmate LSN45) with convertor (DGH D1101). Measured data i.e. total time (t), flow and total volume (V) of filtered water, were recorded by computer with a specially designed software program (Hypfilt Kiwa N.V.) at 5 minute intervals.

The software program calculated the MFI-UF value according to Equation 5.11 as the gradient of two data points with correction to the standard reference conditions of the existing $MFI_{0.45}$ namely; temperature of $20\,^\circ$C, trans membrane pressure of 2 bar (ΔP_o) and surface area of a MFI $0.45\,\mu$m microfilter (A_o):

$$MFI-UF = \frac{\eta_{20^\circ C}}{\eta_T}\ \frac{\Delta P}{\Delta P_o}\ \left(\frac{A}{A_o}\right)^2\ \frac{d\frac{t}{V}}{dV} \qquad \text{...(5.11)}$$

and is therefore directly comparable with the $MFI_{0.45}$. The MFI-UF value was plotted over time for each membrane. The final or real MFI-UF value reported for a membrane indicated as MFI-UF* represents the average calculated from the most stable region of the MFI-UF over time plot in a single test.

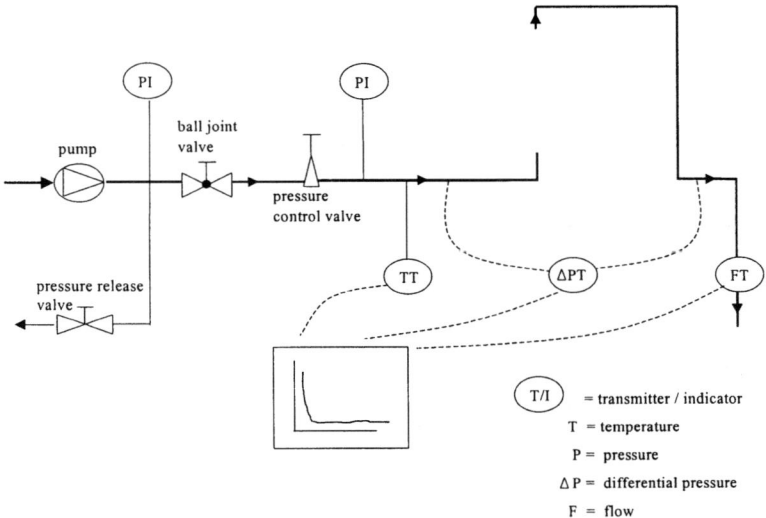

Figure 5.1:   MFI-UF Equipment

5.3.3   Membrane Cleaning

The UF membranes were cleaned after each MFI-UF test. PAN and PS B membranes were backwashed with RO permeate, applied from the permeate outlet, at ambient temperature at a pressure of 1 bar for 15 minutes. Backwashing was not recommended by the manufacturer for PS A membranes. The first chemical cleaning applied for all the membranes was sodium hypochlorite (500 ppm) which was re-circulated for 1 hour at 1 bar at ambient temperature for PAN and PS B membranes and at 60°C for PS A membranes. To restore the clean water flux of the PS A and PS B membranes, several prolonged cleanings with the following solutions; sodium hydroxide (pH 12.5-13.0) up to 60°C (PS A membranes) and citric acid (1%) were required. All cleaning solutions were prepared with analytical grade reagents and RO permeate. To remove residual chemicals, the membranes were backwashed again with RO permeate before the clean water flux was measured.

5.3.4   Membrane Characterisation

5.3.4.1   FESEM

Three clean (unused) hollow fibres were randomly selected from a single membrane, cut open and triplicate samples taken from along the length of the fibre. Samples were mounted on specimen stubbs with the internal surface exposed for coating of the membrane surface. Samples were then chromium coated, circa 2 nm in thickness, using a sputtering coater (Xenosput 2000 Dynavac, Australia) and scanned on the S-900 FESEM (Hitachi, Japan) at an accelerating voltage of 2kV and at magnifications of up to 100 000.

5.3.4.2   Image Analysis

Adjustment of brightness and contrast (Adobe Photoshop 5.0) was carried out on digitised FESEM images to enhance pore details prior to image analysis using Quantimet 500 software (Leica, UK). Due to the irregularity and roughness of the membrane surface, pores were detected and traced manually in Quantimet. Quantimet image analysis calculated individual pore data (area, length, breadth), surface porosity and pore count of the traced pores for each field. Membrane surface characterisation parameters chosen from the Quantimet file were pore area, length and breadth and surface porosity. Pore size distribution data were calculated in Excel for the polysulphone membranes from manufacturer A from the Quanitmet data of pore area.

5.3.5   Filtration Data Analysis

To determine the exponent n in Equation 5.9, which relates the second derivative to the reciprocal flow rate dt/dV, a non linear regression was first performed to smooth the cumulative filtered time and volume data from Hypfilt. This was achieved using the Quasi Newton estimation method in a software program (Statistica 5.0) to fit the following equation:

$$t = a + bV + cV^2 + dV^e \qquad\qquad ...(5.12)$$

where a, b, c, d and e are variables. Subsequently, the second derivative of this equation was calculated using the variables fitted by Statistica 5.0.

To construct the filtration plots of Exp (t) vs. V, t/V vs. V, dt/dV vs. t and t/V vs V to identify complete, standard, intermediate and cake filtration, respectively, the raw cumulative filtered time and volume data from Hypfilt were used directly.

5.4   Results and Discussion

5.4.1   Effect of Membrane Material on the Stability of the MFI-UF Value

The MFI-UF values calculated during filtration tests with tap water are presented in Figure 5.2 for the polyacrylonitrile membranes and the higher (10-100 kDa) MWCO polysulphone membranes and in Figure 5.3 for the lower (1-6 kDa) MWCO polysulphone membranes.

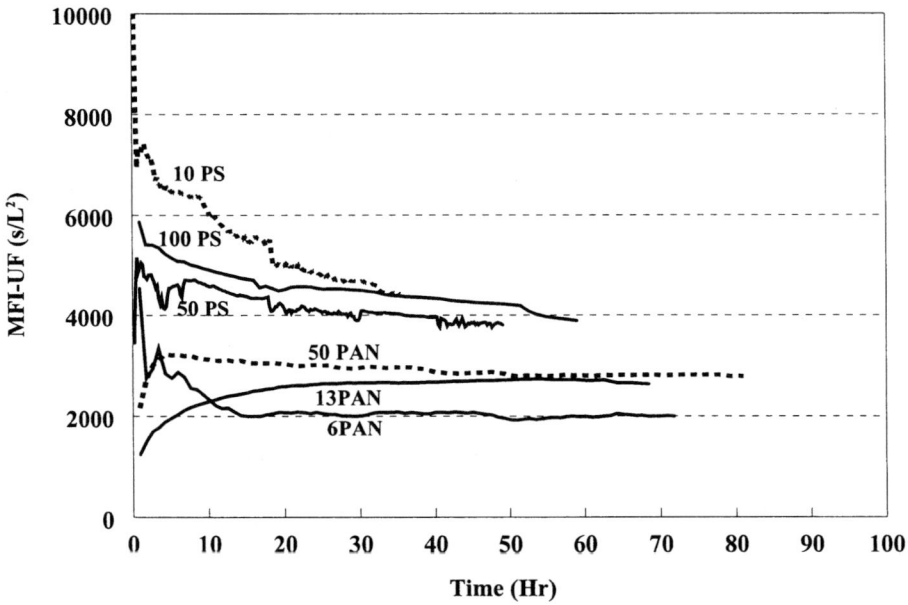

Figure 5.2:   MFI-UF value measured during tap water filtration for polyacrylonitrile (PAN) 6, 13 and 50 kDa membranes and polysulphone (PS) 10, 50 and 100 kDa membranes from manufacturer A.

The MFI-UF* is based on the assumption that cake filtration will occur during the test which results in a minimum or a constant (stable) MFI-UF* value when measured over time. In the case of the PAN membranes a stable region in the MFI-UF versus time plot was observed for all three MWCO membranes after 20-50 hours of filtration until the end of the test i.e. up to 80 hours (Figure 5.2). The MFI-UF versus time plot of the PAN 6 kDa membrane demonstrated the expected behaviour, whereby, the measured MFI-UF was initially high, which corresponds to blocking filtration and thereafter stabilised. In contrast, the MFI-UF value measured for the PAN 50 and 13 kDa membranes was initially low in this experiment and then increased for a period of time (Figure 5.2). This suggests a decrease in flow occurred but is more likely an artifact caused by air trapped in the membrane which partly covered membrane pores and reduced the available filtration area.

Although, the profile of the MFI-UF over time for the polysulphone membranes from both manufacturers was similar to that of the PAN 6 kDa membrane it differed in two main respects. Firstly, the initial MFI-UF value measured was *very* high indicating more severe blocking than with the polyacrylonitrile membranes. Secondly, it continuously decreased over time *approaching* a stable value in some cases after 30-65 hours but in most cases it did not stabilise at all in this period. This latter effect was particularly dramatic for the lower MWCO membranes in the 1-5 kDa range, notably the 1 and 5 kDa membranes which continuously decreased over the entire filtration period up to 100 hours (Figure 5.3).

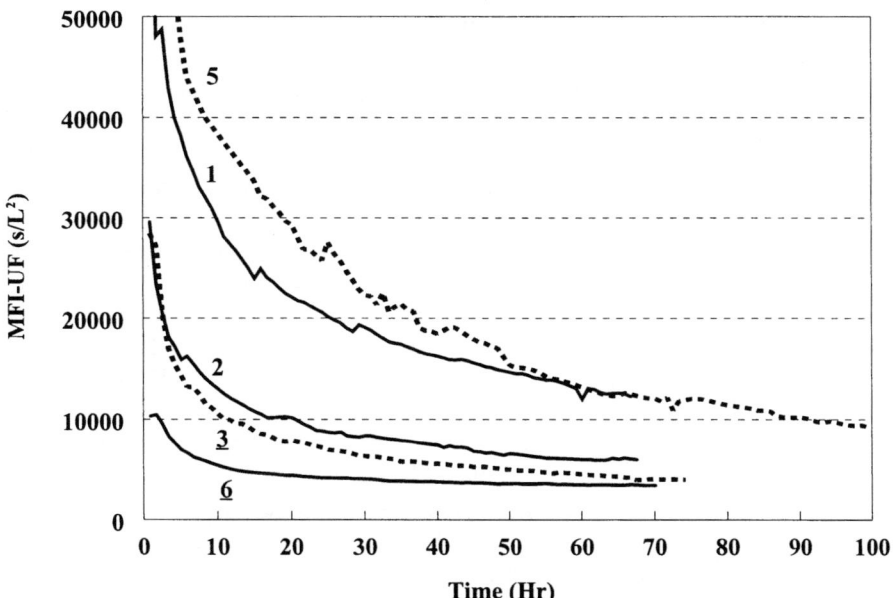

Figure 5.3:   MFI-UF value measured during tap water filtration for the polysulphone 1, 2, <u>3</u>, 5 and <u>6</u> kDa membranes from manufacturer A and <u>B</u>.

The different behaviour observed in the MFI-UF versus time plots *with prolonged filtration* for the two membrane materials was not expected to be due to differences in membrane charge or hydrophobicity. Although, the membrane charge was not quantified in this research, both membrane materials are expected to be negatively charged. Polysulphone membranes have been shown in numerous studies to be negatively charged [24-26]. Polyacrylonitrile membranes can also be assumed to have a negative charge, as membranes manufactured by the phase inversion process typically bear a negative charge [27]. Only an *initial* delay in forming a cake layer on both the polyacrylonitrile and polysulphone membranes may have occurred due to electrostatic repulsion forces between the colloids, typically negatively charged in natural conditions, and the membrane. This is because the MFI-UF test is applied in dead end mode and this should minimise charge effects on the MFI-UF value, especially with such long filtration run times. Similarly, differences in the hydrophobicity of the two membrane materials probably only affect initial cake layers as to whether the particles are chemically adsorbed onto the membrane or simply deposited on the membrane surface.

A factor determining the onset of cake filtration may be the magnitude of the initial clean water flux. Higher flux membranes will bring particulate material to the membrane surface at a faster rate. Cake build up may then be faster and therefore the MFI-UF value might stabilise earlier. However, plots of the MFI-UF value versus filtered volume for both polysulphone and polyacrylonitrile membranes (not shown) were not significantly different to the MFI versus time plots of Figure 5.2 and 5.3. Thus, the MFI-UF value for the polysulphone A and B series was found to decrease with both filtered volume and filtration time and no constant (stable) MFI-UF* value was obtained. Conversely, for the PAN membranes the MFI-UF value stabilised after either sufficient filtered time or filtered volume. No trend between the initial clean water flux and when or if the MFI-UF value stabilised was found for either membrane materials. For instance the clean water fluxes of the PS 2 and 10 kDa membranes were similar, yet, the 10 kDa membrane was comparatively more stable over time than the 2 kDa membrane. Similarly, the flux of the PAN 6 kDa membrane was 7 times lower than the PAN 50 kDa membrane and yet it gave a stable MFI-UF value after 30 hours, while the latter membrane required 50 hours.

However, the extent of the initial sharp decline observed for the polysulphone membranes from both manufacturers appeared to be related to the initial clean water flux. For example the 2 and 6 kDa membranes have the highest initial fluxes for polysulphone membranes in the 1-6 kDa MWCO range and the initial decline in the MFI-UF value was less severe than for the other polysulphone membranes (refer Figure 5.3). The initial sharp decline in the MFI-UF value most likely corresponds to blocking filtration. Low surface porosity membranes are known to be susceptible to blocking and the loss of pores in such a membrane will result in a significant loss of the membrane flux [10]. Therefore, the general behaviour observed for the polysulphone membranes may be due to differences in membrane surface characteristics such as surface porosity and pore size distribution rather than due to the chemical nature of the material. Although, these factors are expressed in the resistance of the membrane (Equation 5.3) and hence the initial clean water flux via Equation 5.2, one particular factor e.g. surface porosity may play a dominant role in the permeability of the surface and in determining the extent of blocking filtration and when cake filtration occurs for a membrane.

5.4.1.1   Effect of Membrane Material on Cleaning Efficiency

The results of a hydraulic (backwashing) and a chemical cleaning of the polyacrylonitrile and selected polysulphone membranes after an MFI-UF test are summarised in Table 5.3. Membrane hydrophobicity and not charge was expected to determine the efficiency of membrane cleaning. Both the membrane materials and the particles are expected to be negatively charged as discussed earlier and thus electrostatically repel each other. Thus, backwashing should easily remove the particles. However, hydrophobic membranes are widely reported to be more susceptible to adsorptive fouling by organic particles [24,28-32]. As polysulphone is reported to be more hydrophobic than polyacrylonitrile membranes [32,33] the success of backwashing these membranes was expected to be limited.

This expectation was supported by the results as backwashing gave only 30-48% clean water flux restoration for the polysulphone B membranes. Whereas, backwashing of the polyacrylonitrile membranes from the same manufacturer was almost sufficient to clean the membranes, yielding >90% of the initial clean water flux and a single chemical cleaning fully restored the flux. Therefore, the fouling of the polyacrylonitrile membranes was more likely dominated by physical deposition. Whereas in the case of the polysulphone membranes the particles were more likely chemically adsorbed onto the membrane surface as a single chemical cleaning, increased the clean water flux restoration up to 75%. Moreover, total restoration of the clean water flux of these membranes and the other polysulphone membranes from manufacturer A (refer Table 5.3), required 3-6 chemical cleanings. Thus, in terms of membrane reuse the polyacrylonitrile membranes appeared to be more suitable for the MFI-UF test.

Table 5.3: Effect of a single backwashing and a chemical cleaning on clean water flux restoration for selected polysulphone and polyacrylonitrile membranes.

Membrane Material	MWCO kDa	% Clean Water Flux Restoration	
		after 1 backwash	after 1 chemical cleaning
Polysulphone manufacturer A	1	*	50
	5	*	35
	10	*	44
	100	*	50
Polysulphone manufacturer B	3	30	60
	6	48	75
Polyacrylonitrile manufacturer B	6	95	99
	13	91	99
	50	90	99

(*) backwashing not recommended by the manufacturer for these membranes.

5.4.2   Effect of Membrane MWCO on the MFI-UF* Value

The MFI-UF* value measured for tap water by the three membrane series are summarised in Table 5.4. The corresponding filtered time and volume from where the MFI-UF* value was determined is included in Table 5.4. The MFI-UF* value reported for the polysulphone membranes, particularly PS 1 and 5 kDa, was determined as an average from the last portion of the filtration test despite the lack of stability in the MFI-UF value over time. It was assumed that the tap water did not change over the experimental period as the MFI-UF* value measured in duplicate experiments for selected polysulphone membranes (series A) gave similar results (refer Figure 5.4).

Table 5.4:   MFI-UF* value determined in tap water experiments with the corresponding filtered time (A) and filtered volume (B) range using polysulphone and polyacrylonitrile membranes of 1-100 kDa MWCO.

Membrane Material	MWCO kDa	MFI-UF* (s/l^2) determined from the *most* stable MFI-UF region with the corresponding filtered time (A) and volume (B) range		
			(A) hour	(B) Litres
Polysulphone manufacturer A	1	13 300[1]±60	50-70	370-470
	2	6000[1]±180	55-70	870-1010
	5	8400[1]±600	60-110	460-760
	10	4500	30-40	690-730
	50	3800[1]±81	30-50	840-1120
	100	4000	50-60	230-250
Polysulphone manufacturer B	3	4100	65-74	356-400
	6	3500	50-70	406-520
Polyacrylonitrile manufacturer B	6	2000	30-70	240-500
	13	2700	20-70	340-724
	50	2800	50-80	630-860

(1) average and standard deviation of measured MFI-UF* in duplicate experiments.

The MFI-UF* value ranged from 2000 to 13 300 s/L^2. In comparison, the $MFI_{0.45}$ commonly measured for tap water in the Netherlands is much lower, in the range of 1-5 s/L^2. A higher MFI-UF* value suggests the retention of smaller particles which increases the specific resistance of the cake and the MFI-UF* value via the Carmen-Kozeny Equation (Equation 5.5). The high MFI-UF* value obtained for tap water also demonstrates that particles remain in the tap water even after conventional pretreatment which may foul UF membranes. Moreover, these particles could not previously be measured by the existing $MFI_{0.45}$ (and SDI) fouling indices.

The effect of membrane MWCO (1-100 kDa) on the MFI-UF* value determined for tap water by the three membrane series is presented in Figure 5.4. Assuming a decrease in MWCO corresponds to a decrease in pore size, a higher MFI-UF* value was expected for lower MWCO membranes due to the retention of smaller colloids by the membrane. However, only the

polysulphone membranes in the 1-5 kDa MWCO range from manufacturer A gave very high
MFI-UF* values of 8400 - 13 300 s/L². The other membranes ranging in MWCO from 3 to 100
kDa gave markedly lower MFI-UF* values of 2 000 to 4500 s/L². Moreover, the MFI-UF* values
for these membranes did not appear to be related to the membrane MWCO.

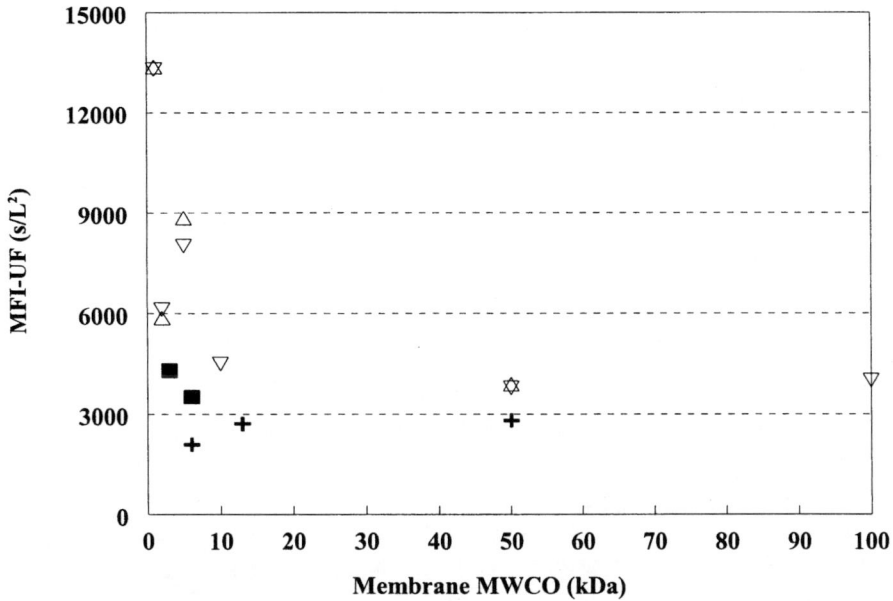

Figure 5.4: Effect of membrane MWCO on the measured MFI-UF* value obtained for tap
water for the polyacrylonitrile membrane series (✚) and polysulphone membrane
series; manufacturer A ((△) duplicate measurement (▽)) and manufacturer B (■).

The results observed may be attributable to different methods applied by the manufacturers in
rating MWCO. However, for the polysulphone series from manufacturer A, where it can be
assumed the same method for determining MWCO was applied, two distinct regions of MFI-UF*
dependence on MWCO were evident. In the first region corresponding to 1-5 kDa, the MFI-UF*
was found to be MWCO dependent and in the second region, MWCO independent, in this latter
region a tenfold increase from 10 to 100 kDa gave only a ≈13% decrease in the MFI-UF* value
(Table 5.4). Furthermore, a comparison of the PAN 6 and PS 6 kDa membranes from the *same*
manufacturer with the same MWCO rating were expected to give similar MFI-UF* values.
Notwithstanding, the MFI-UF* value measured for the PS 6 kDa was significantly higher than
the PAN 6 kDa membrane. This may be due to a lower surface porosity of the PS 6 kDa
membrane, reducing the filtration area in Equation 5.8 which increases the MFI-UF value. Thus,
higher MFI-UF* values measured, may not in fact correspond to the retention of smaller
colloidal particles but as a result of surface properties such as surface porosity, which further
emphasises the need for an alternative method for membrane characterisation.

5.4.3   Membrane Surface Examination by FESEM

5.4.3.1   Polyacrylonitrile Membranes

FESEM images of the three MWCO polyacrylonitrile membranes revealed the surfaces to be composed of fine smooth fibres (Figure 5.5A-C) with pores located between the fibres. Image analysis of the polyacrylonitrile membrane surfaces to determine surface porosity and pore dimensions are reported in Table 5.5. However, most likely the real surface porosity is higher as some pores may have been obscured by the fibres which hindered their detection. Nevertheless, despite the low surface porosity (2.0-3.5%) of the membranes detected, they were assumed to be homogeneously permeable, as the fibres were estimated to be circa 15-20 nm in width and in a regular repeating arrangement. A homogeneously permeable membrane probably led to a more even cake deposition over the membrane surface and hence explains the more stable MFI-UF value observed for these membranes after sufficient filtered time (refer section 5.4.1). This will be discussed in section 5.4.4. in more detail.

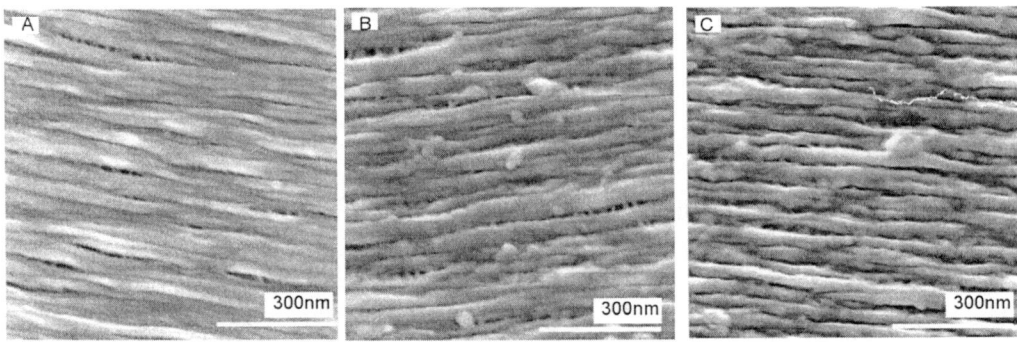

Figure 5.5:   FESEM Micrographs of the polyacrylonitrile 6 kDa (A), 13 kDa (B) and 50 kDa (C) membranes (×100 000 magnification).

The MFI-UF* value increased in the order PAN 6 kDa < 13 kDa ≈ 50 kDa. FESEM results in Table 5.5 indicate that the 6 kDa membrane has on average larger pore dimensions and a higher surface porosity than the 13 kDa membrane. Both these membrane surface characteristics would lead to a lower MFI-UF* value; a higher surface porosity would increase the filtration area (Equation 5.8) while larger pores would only retain larger particles which have less effect on the specific resistance (Equation 5.5). However, the membrane resistance of 6 kDa (included in Table 5.5) calculated from the clean water flux, was significantly higher than the 13 kDa membrane and contradicts the FESEM findings. Membrane resistance quantifies the permeability of the 6 kDa membrane and is dependent on surface porosity (assuming straight through pores), tortuosity, pore size and membrane thickness via Equation 5.3. Membrane thickness was estimated from FESEM images to be equal for all three polyacrylonitrile membranes at ≈20 µm. Tortuosity of the pores may have increased membrane resistance but only to a limited extent. The higher membrane resistance of the 6 kDa membrane may be attributable to a lower surface porosity and/or a smaller pore size than that determined by FESEM. Alternatively and more likely, the pores are not straight through pores but the vary with depth. The pores of the 6 kDa membrane may have tapering pores which decrease significantly with depth. Whereas, the higher surface porosity and larger pore size FESEM results for the 50kDa membrane agree with the lower membrane resistance calculated (refer Table 5.5) and suggest the pores of the 50 kDa membranes may be more constant with depth.

Table 5.5: Membrane resistance calculated from clean water flux and membrane pore size characteristics and surface porosities as determined by FESEM.

Membrane	MWCO kDa	Membrane Resistance (m^{-1})	Pore Area (nm^2)	Pore Length (nm)	Pore Breadth (nm)	Surface Porosity (%)
Polysulphone manufacturer A	1^1	5.3×10^{12}	215 ± 92	21 ± 5	13 ± 3	2.9 ± 0.9
	2^1	2.1×10^{12}	172 ± 79	22 ± 7	10 ± 3	3.2 ± 1.2
	5^1	3.8×10^{12}	220 ± 93	22 ± 6	13 ± 3	3.1 ± 0.9
	10	2.1×10^{12}	231 ± 10	22 ± 6	13 ± 4	4.1 ± 1.5
	50	1.2×10^{12}	246 ± 12	24 ± 8	13 ± 4	3.5 ± 1.0
	100	1.4×10^{12}	220 ± 93	22 ± 6	13 ± 3	5.2 ± 0.8
Polysulphone manufacturer B	3	3.8×10^{12}	-	-	-	-
	6	1.8×10^{12}	204 ± 96	24 ± 7	11 ± 3	2.0 ± 0.4
Polyacrylonitrile manufacturer B	6	6.5×10^{12}	194 ± 86	24 ± 8	11 ± 3	2.6 ± 0.5
	13	2.1×10^{12}	150 ± 87	21 ± 15	9 ± 3	1.9 ± 0.5
	50	8.9×10^{11}	214 ± 87	27 ± 8	10 ± 2	3.5 ± 0.9

(*[1]*) *measurements from striations on the membrane surface and represent maximum surface porosity and pore dimensions of the striated area only.*

Based on the FESEM ranking of increasing pore size and surface porosity; 13 kDa < 6 kDa < 50 kDa it could be expected that the measured MFI-UF* should be 13 kDa > 6 kDa>50 kDa. This was not the case. Therefore, the FESEM results cannot fully account for the MFI-UF values measured with the PAN membrane series. A possible explanation for these results is compression of the cake on the membrane surface caused by the significantly higher fluxes of the 13 and 50 kDa membranes which are 3-7 times higher than for the 6 kDa membrane. Compression of the cake would increase the specific cake resistance and hence the MFI-UF* value. The effect of cake compression will be discussed in Chapter 6.

5.4.3.2   Polysulphone Membranes

Low magnification FESEMS of polysulphone A membranes in the 1-5 kDa MWCO range, revealed the presence of striations running lengthwise across the surface. This is illustrated in Figure 5.6A (×1000 magnification) for the 1 kDa membrane and is also representative of the surfaces observed for the 2 and 5 kDa membranes. At higher magnification (×5000 and ×30 000), the striated surfaces of these membranes appear more rough in appearance (refer Figure 5.6B and 5.6C). Pores were principally found on or close to these rough striations as shown in Figure 5.6D and 5.6E, respectively. Whereas, between the striations, no pores were observed (Figure 5.6F) and the surface of these non porous bands appeared to be very smooth. Moreover, the non porous bands were not constant in size but were present as very wide and very narrow bands (refer Figure 5.6A). Consequently, these membrane surfaces are heterogeneously permeable with flow through these membranes limited to the narrow striated porous regions.

Surface porosity and pore size dimensions determined for the striated (porous) regions of the 1-5 kDa membranes, are presented in Table 5.5. It was not possible to ascertain the proportion of membrane area covered with striations. However, if the striated or porous region was averaged over the entire membrane surface, the membrane surface would be of a very low surface porosity. This equates to a very low filtration area in Equation 5.8 which explains the very high MFI-UF* values found for the 1-5 kDa polysulphone A membranes. The anomalously high MFI-UF* value of the 5 kDa membrane is mostly likely as a consequence of the difficulty in manufacturing striations on these membranes reproducibly. This is supported by the higher membrane resistance calculated from the clean water flux for this membrane (Table 5.5) which indicates that it is structurally more similar to the 1 kDa than the 2 kDa membrane. The 1 and 5 kDa membranes may have less striations and hence lower "average" surface porosity than the 2 kDa membrane resulting in their higher MFI-UF* values.

No striations were found on the surfaces of the 10, 50, and 100 kDa membranes from manufacturer A or the 3 and 6 kDa membranes from manufacturer B, pores were found equally distributed over the entire surface (refer Figures 5.7 and 5.8, respectively). At higher magnification, membranes from the latter manufacturer appeared to be smoother than the former. In particular, the 3 kDa (manufacturer B) appeared to be a very tight membrane of low surface porosity (refer Figure 5.8C) which is supported by the higher membrane resistance in comparison to the 6 kDa. Pores could not be visualised with any accuracy for the 3 kDa membrane and therefore no image analysis was carried out. The surface porosity and pore dimensions; area, length and breadth determined for the other membranes are included in Table 5.5.

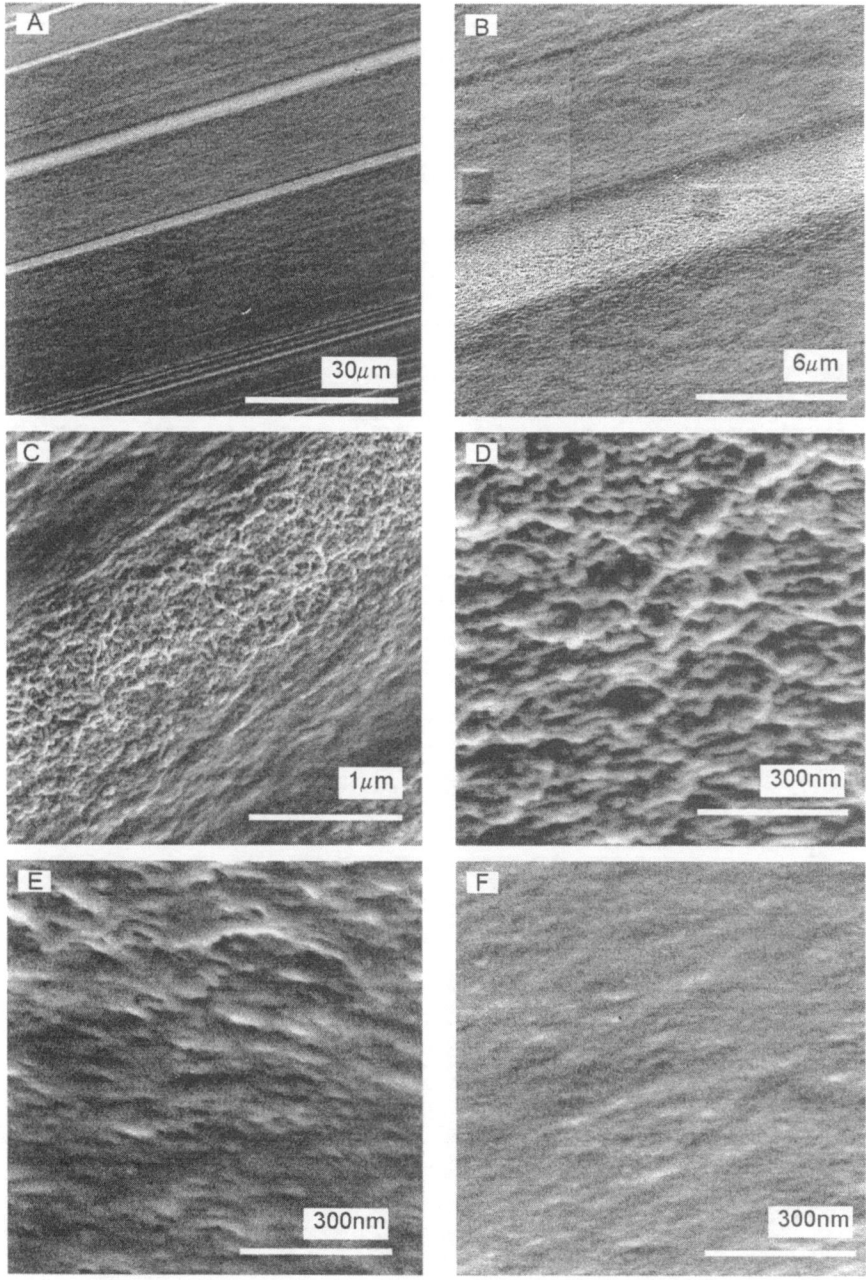

Figure 5.6: FESEM micrographs of the polysulphone 1 kDa membrane at low magnification ×1000 (A) showing the presence of striations across the membrane surface. At higher magnification ×5000 (B) and ×30 000 (C) the striations are shown to vary in roughness due to the presence of pores, visible at ×100 000 magnification, on or close to a striation (D) and (E) and absence (F) of pores between the striations. The location of (D) and (F) can be observed as the raised areas in (B).

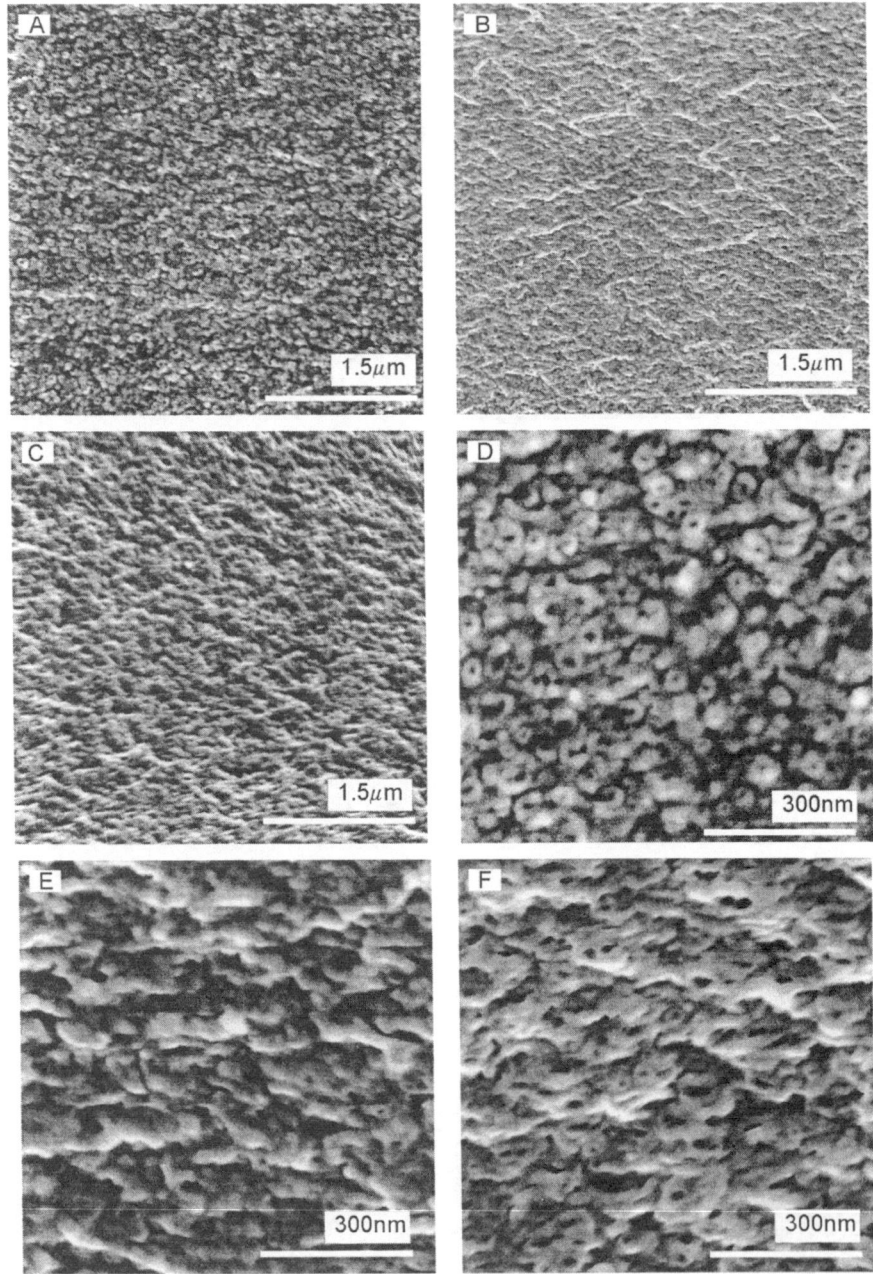

Figure 5.7: FESEM micrographs of the polysulphone 10 (A and D), 50 (B and E) and 100 (C and F) kDa membranes showing the homogenously porous surface at low ×20 000, and high ×100 000 magnification, respectively.

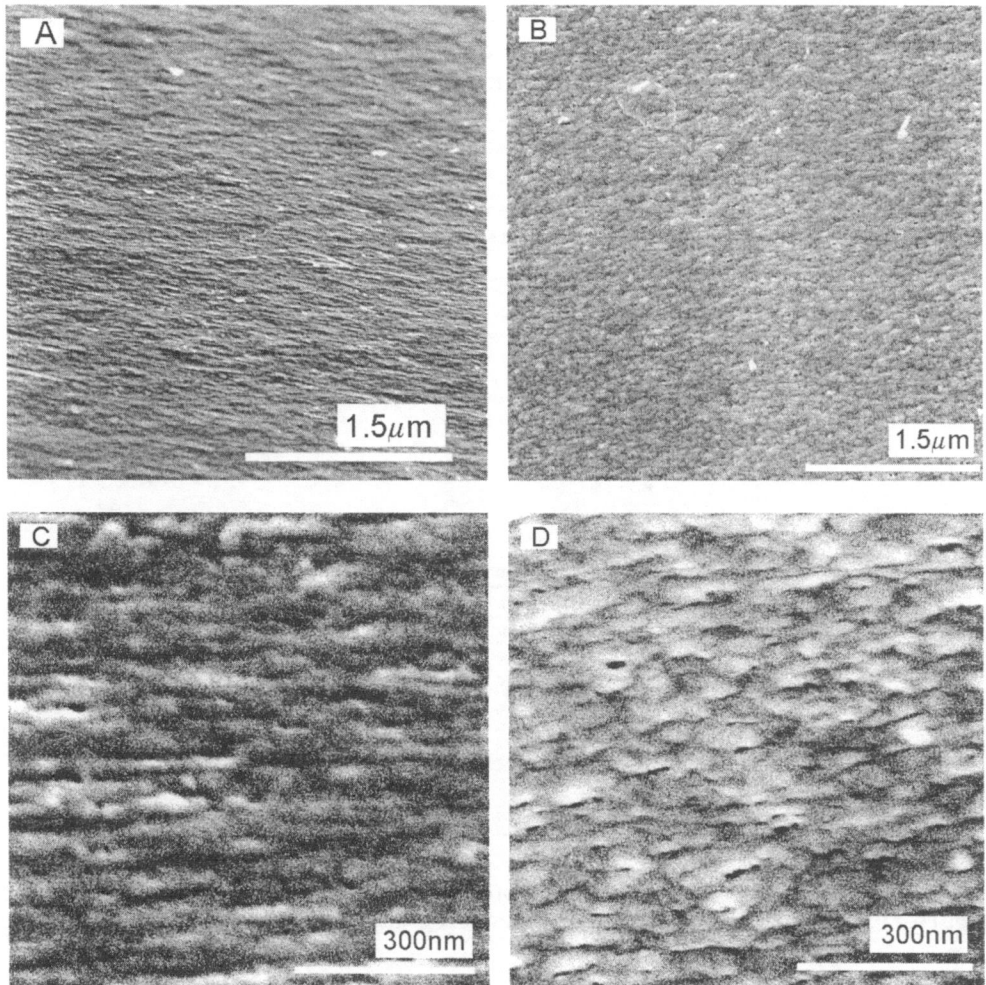

Figure 5.8:   FESEM micrographs of the polysulphone 3 kDa (A and D), and 6 kda (B and D) membranes from manufacturer B showing the surface at low ×20 000 and at high ×100 000 magnification.

The surface porosity and pore size dimensions for the 10 - 100 kDa membranes and for the striated regions of the 1-5 kDa membranes (A series) were found to be surprisingly similar for membranes ranging in MWCO from 1-100 kDa. This was also evident from the pore area distribution of these membranes, plotted in Figure 5.9, where the highest frequency of pores were found to be in the range of 100-300 nm^2. Ignoring the striations of the 1, 2 and 5 kDa membranes, the only differences observed for these membranes was a higher frequency of pores in the 100-200 nm^2 range. Thus, despite up to a 100 fold increase in MWCO no pronounced difference in pore size was observed. However, as discussed previously for the polyacrylonitrile membranes the MWCO rating may be due a smaller pore size inside the membranes than in the outer skin.

Although, the 1-100 kDa membranes are very similar on the basis of their pore dimensions the striated porous regions and non porous bands of the 1-5 kDa membranes may alter the permeation behaviour of these membranes. Furthermore, the MFI-UF* values of the 1-5 kDa membranes are most likely artificially high due to the reduced porous area over their membrane surfaces. If the filtration area in Equation 5.8 could be corrected for the limited porous regions of these membranes most likely the MFI-UF* values would be similar to that of the 10, 50 and 100 kDa membranes. For instance if these membranes were 4 times less porous than the other polysulphone membranes from this series the MFI-UF* would be 16 times greater via Equation 5.8.

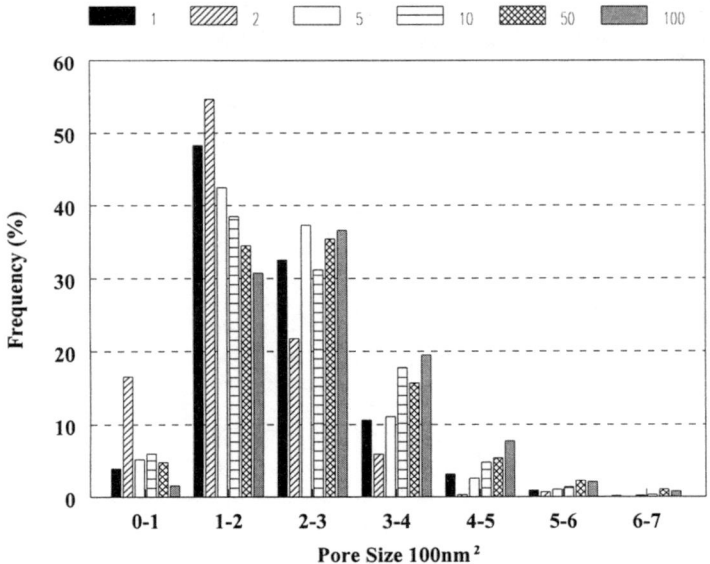

Figure 5.9:   Comparison of pore area distribution for the polysulphone membrane series from manufacturer A.

5.4.4   Analysis of Filtration Mechanisms

Attempts to identify the filtration mechanisms occurring during an MFI-UF test by determining the exponent n from plots of the second derivative d^2t/dv^2 vs. dt/dV using the smoothed t and V data proved inconclusive for both the PAN and PS membranes. This was attributed to disturbances in the manual operation of pressure valves as the system stabilised at the start of filtration. While towards the end of filtration, fluctuations in the raw data increased due to the sensitivity of the flow meter as the flow decreased over time. In contrast, the first derivative or integrated equations of the individual filtration models proposed by Hermans and Bredée [19] and Hermia [21] proved more successful as errors become smaller and were effectively absorbed as the cumulative t and V data increased as filtration proceeded. Consequently, these models were applied to identify the sequence and dominance of filtration mechanisms occurring during tap water filtration for the PAN and PS membranes and are presented in Figures 5.10 and 5.11, respectively.

Intuitively, complete (Plots 5.10A and 5.11A) and standard blocking (Plots 5.10B and 5.11B) were expected to occur at the beginning of filtration as particles completely seal smaller pores and progressively fill larger pores. Intermediate blocking (Plots 5.10C and 5.11C) was expected to occur consecutively for an extended period of time as a transition phase between blocking (complete and standard) and cake filtration i.e. until a significant cake layer had build up. However, no clear separation in time was found for these blocking mechanisms in the relevant plots. Instead linearity was observed from the start of filtration up to 5 hours in all three blocking plots for all membranes, except the PAN 6 and 13 kDa. This suggests that complete, standard and intermediate blocking occurred simultaneously. For the PAN 6 and 13 kDa membranes, the intermediate blocking phase was extended up to 15 hours (refer Figure 5.10C).

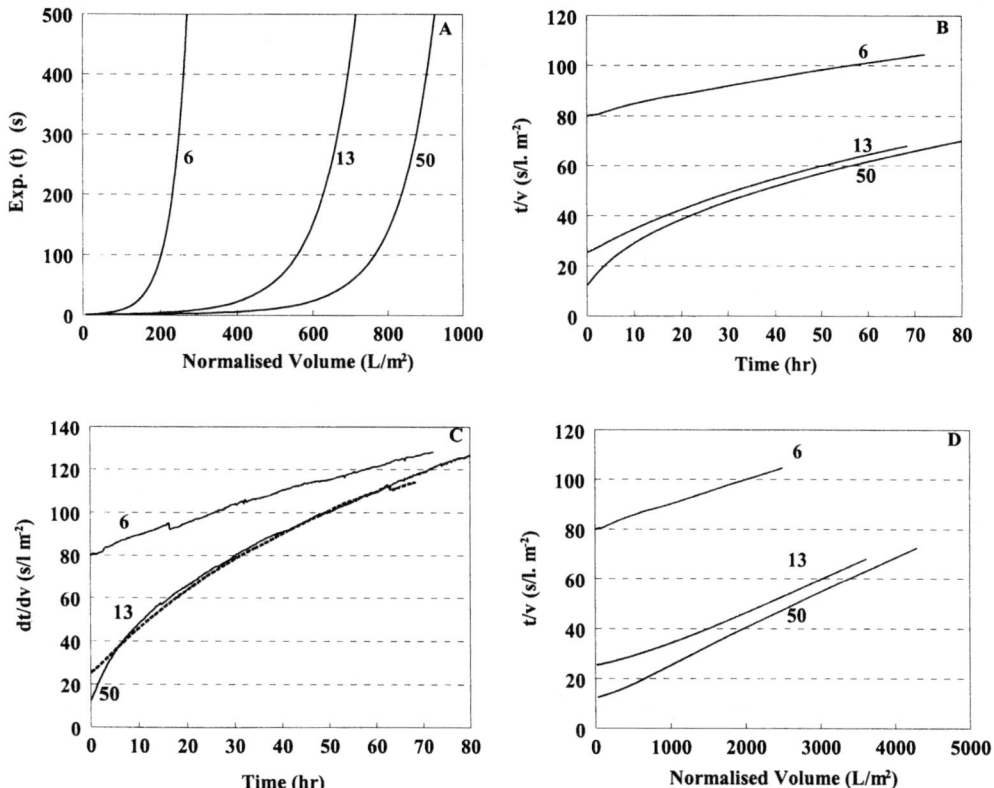

Figure 5.10: Comparison of filtration mechanisms for the polyacrylonitrile membranes: complete blocking (A), standard blocking (B), intermediate blocking (C) and cake filtration (D).

It has to be noted that a precise estimation of the intermediate and standard blocking phases proved difficult so the aforementioned time periods are rather arbitrary. For instance in the standard blocking plot of the PAN 6 kDa membrane, linearity was observed up to 3 hours initially and later between 20-50 hours of filtration. In addition, the linearity observed was influenced by the scale applied to the axes as demonstrated in the intermediate plots for the polysulphone membranes (refer Figure 5.11C). In these plots linearity occurred briefly, in the order of 2-5 hours and for the PS 2, 3 and 6 kDa membranes the data appeared to approach linearity once again after 30 hours. Although, on scale expansion of this area (not shown) these plots were not found to be linear. Only the extent of complete blocking could be distinguished with any certainty, with a clear trend evident in the Exp (t) versus V plots (5.10A and 5.11A) where the higher MWCO PS and PAN membranes required a higher volume before deviating from linearity. This could be correlated with the permeability of the membrane (see membrane resistance Table 5.5). For example in the polyacrylonitrile series, circa 100, 300, 500 litres, was required for complete blocking of the PAN 6, 13 and 50 kDa membranes, respectively.

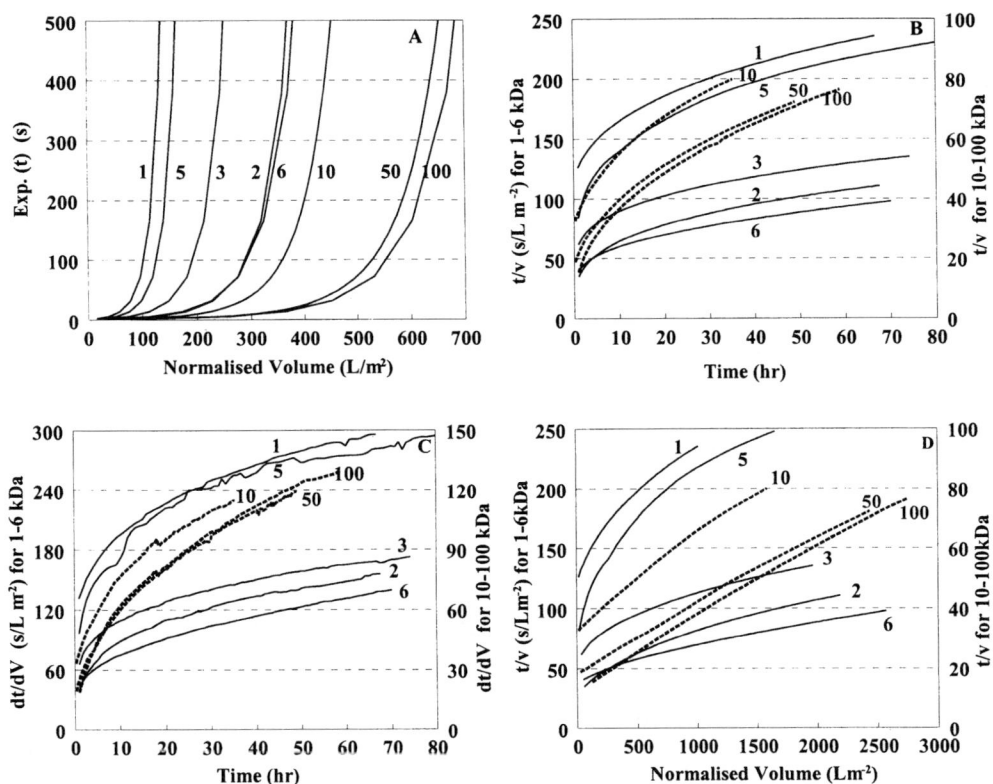

Figure 5.11: Comparison of filtration mechanisms for the polysulphone membranes: complete blocking (A), standard blocking (B), intermediate blocking (C) and cake filtration (D).

Cake filtration was expected to occur following blocking filtration. As the MFI-UF test is based on cake filtration this filtration mechanism must be demonstrated. Linearity in the t/V versus V plot (Plots 5.10D and 5.11D) indicates cake filtration which results in a stable or minimum MFI-UF value when plotted over time. Proof of cake filtration for all three polyacrylonitrile membranes was found in Figure 5.10D after 1200, 1700 and 3100 litres filtered volume, respectively which coincides to the region in Figure 5.2 where the MFI-UF stabilised over time.

In contrast, no linear relation was found in the t/V versus V plot for the polysulphone membranes (refer Figure 5.11D). Although, the higher MWCO 10-100 kDa polysulphone membranes *appeared* to be linear in the t/V versus V plot, as observed with the blocking filtration plots, linearity was influenced by the choice of scales. The lack of linearity in the data for the higher MWCO polysulphone membranes was more readily visible in the MFI-UF plot over time (Figure 5.2 and 5.3) as it is the gradient of the t/V versus V plot.

When the sequence of filtration mechanisms over the entire filtration period was examined, gaps were found for all membranes where the exact filtration mechanism could not be ascertained with any certainty at all. The unaccounted time period was most pronounced for the polysulphone membranes where the various blocking mechanisms could be detected up to 2-4 hours after which no filtration mechanism appeared to prevail. While this gap was smallest for the PAN 13 kDa membrane, as intermediate blocking occurred up to 15 hours and cake filtration was established at 20 hours. The difficulty in establishing the extent and sequence of filtration mechanisms may be attributable to the simplifying assumptions of the various models which limits their ability to describe such complicated phenomena and in distinguishing between them when they occur simultaneously. For instance complete blocking assumes that each pore is blocked by one particle, whereas, in reality a large particle may block either multiplet pores or two or more pores in a high surface porosity membrane. Moreover, with the exception of intermediate blocking, the models were developed describing the exclusive occurrence of one mechanism. Although, intermediate blocking is less restrictive being based on complete pore blocking combined with cake filtration, it still doesn't take into account the occurrence of pore constriction or standard blocking.

Recent studies by Ho and Zydney [34,35] highlighted the limitations of applying the classic blocking mechanisms to polymeric microfiltration membranes with a highly interconnected structure (isotropic) e.g. polyethersulphone and mixed cellulose ester membranes. These membranes typically have an open fiberous network structure of interconnected pores. Thus, as discussed by Ho and Zydney [35] the underlying assumption of the blocking models that the membrane consists of an array of parallel capillary tubes of constant diameter and length (i.e. the idealised membrane with straight through pores) is invalid. A much lower rate of flux decline was demonstrated by Ho and Zydney for an isotropic membrane compared to that of a membrane of similar porosity but with straight through non connecting pores [34]. Ho and Zydney [34,35] explained that particle deposition on the upper surface of an isotropic membrane which blocked pores, affected the overall resistance to a much lesser extent due to a change in the fluid flow profile as the fluid could flow through the interconnected pore structure under and around the blockage. This effect was shown to depend on membrane thickness. In their analysis of protein fouling through asymmetric polyethersulphone membranes having a very thin skin layer of $\approx 0.5\mu m$, the membrane behaved as it had straight through pores [34]. Ho and Zydney developed a model to account for effects of surface pore blockage for isotropic membranes. Although, the membrane thickness of the PS and PAN membranes used in this research was estimated at 15-20μm, the FESEM images of the surface do not appear mesh like in structure but are smother in appearance than the FESEM images of the membranes used by Ho and Zydney [34]. Moreover, a certain degree of internal blocking was expected for the membranes used in this research rather than the surface deposits as modelled by Ho and Zydney [35]. In addition, the filtration times in this research are long and a cake was expected, although not demonstrated for the PS membranes in the cake filtration plot. Whereas, the model developed by Ho and Zydney [35] was for isolated particulate foulants and not for a multilayer cake and therefore, cannot be applied to this data to prove cake filtration.

Assuming that cake filtration did occur for the polysulphone membranes, the changing gradient in the t/V versus V plots may be explained by the effect of lateral flow of the streamlines due to the irregular distribution of pores and/or the low surface porosity of their membrane surfaces. As a consequence of the limited permeability of the polysulphone surface, especially for the striated 1, 2, and 5 kDa membrane surfaces, flow through the membrane will have had a significant lateral component increasing the cake resistance as suggested by Fane [36] and depicted in Figure 5.12. As filtration proceeds, cake build up around the mouth of the pores increases and the effective filtration area increases. The MFI-UF is a function of the filtration area squared and thus surface porosity (Equation 5.8), consequently the MFI-UF decreases over time as the filtration area increases. After sufficient cake build up, surface porosity has a diminished effect and axial flow dominates as a result the cake filtration mechanism will be fully developed. At this point the MFI-UF will stabilise. Therefore, if filtration had continued for a sufficiently long period of time a stable MFI-UF* value would have been obtained for the polysulphone membranes. This mechanism, in particular when the pores are not evenly distributed over the surface, might be responsible for the differences in the required time/filtered volume to reach the stage of stable cake filtration and thus a stable MFI-UF* for the different membranes.

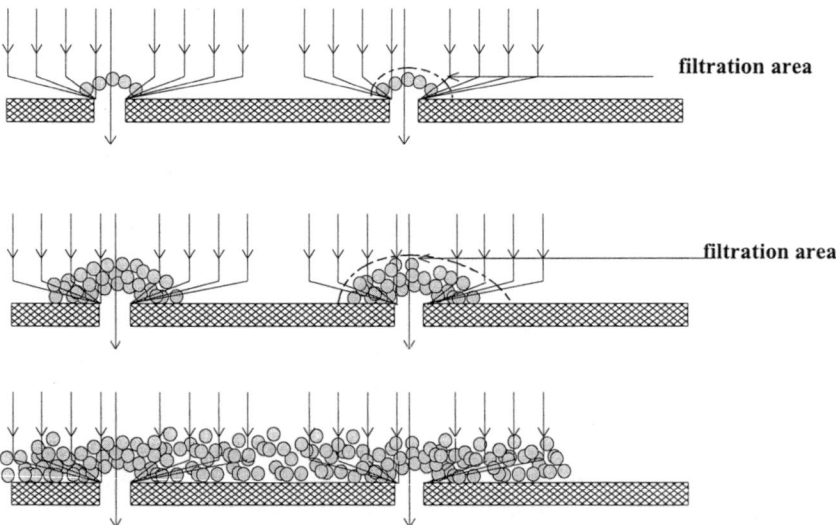

Figure 5.12: Schematic representation of the effect of low membrane surface porosity on cake build up and change in filtration area in the polysulphone membrane series.

Moreover, once surface porosity and/or irregular pore distribution effects can be neglected, most likely the size of the interstices of the cake layer itself takes over the role in determining the size of particles retained and hence the resultant MFI-UF* value obtained. Thus, the cake acts as a second membrane. It is commonly accepted that particles 10 times smaller than the pore size can be retained in a cake. This would explain why the MFI-UF* determined for the PAN and PS membranes appear rather independent of membrane MWCO or pore dimensions in Figure 5.4 and appeared to vary only to a small extent from 2000 to 4500 s/L^2 despite the MWCO varying from 3-100kDa. Exceptions to this trend were the PS 1-5 kDa membranes for which the MFI-UF* was still artificially high due to surface porosity effects as discussed above and the PAN 6 kDa membrane where the low MFI-UF* measured is attributed to its lower flux and cake compressions effects. Thus, if filtration had been prolonged for all the polysulphone membranes and excluding the MFI-UF* value of the PAN 6 kDa, the MFI-UF* might be similar to that obtained for the PAN 13 and 50kDa membranes of 2700-2800 s/L^2.

5.4.5   Proposed Reference UF Membrane for the MFI-UF Test

Of the two membrane materials examined in this study, polyacrylonitrile membranes were found to be more suitable for application in the MFI-UF test due to the following reasons; the MFI-UF* could be determined within 20-50 hours of tap water filtration and cake filtration was proven by linearity in the t/V versus V plot which gave a stable MFI-UF over time. Furthermore, the clean water flux of these membranes could easily be restored by one cleaning with sodium hypochlorite. Whereas, for the polysulphone membranes, cake filtration could not be proven or required excessively long filtration time (>60 - 100 hours of filtration) to obtain the real or stable MFI-UF* value. Moreover, these membranes needed repeated chemical cleanings to restore the clean water flux.

From the polyacrylonitrile membrane series, the 13 kDa membrane was proposed as the most promising reference membrane for use in the MFI-UF test for the following reasons. The filtration mechanisms were more difficult to identify for the PAN 50 and 6 kDa membranes and the MFI-UF* obtained for the latter membrane was expected to be lower due to its lower flux and less compression of the cake. Whereas, the sequence of filtration mechanisms was easier to identify for the PAN 13 kDa membrane with intermediate blocking occurring up to 15 hours of tap water filtration and after 20 hours cake filtration was proven to be the dominant filtration mechanism.

The size of particles present in the tap water and which were captured in the MFI-UF test was not determined in this research. Measurements of the particle size distribution of a RO feedwater after conventional pretreatment and ultrafiltration (150-200 kDa) are presented in Table 5.6. This shows smaller particles are present in higher numbers, with the largest proportion of particles remaining in the feedwater in the smallest channel range measured by the particle counter of 50-100 nm. Therefore, most likely even larger numbers of particles smaller than the detection limit of the counter (50nm) are present. Previously the fouling potential of these smaller particles could not be measured as they most likely passed through the pores of the existing MFI$_{0.45}$ test membrane, calculated at $0.15 \mu m^2$. Whereas, the average pore size estimated by FESEM for the PAN 13 kDa membrane is 1000 times smaller at 150 nm^2. Furthermore, due to the entrapment of particles within the cake as discussed in section 5.4.4. a proportion of these particles may be retained and measured by the MFI-UF test membrane.

Table 5.6:   Particle size distribution of a conventionally pretreated feedwater after further treatment by ultrafiltration (150-200 kDa).

Channel Size Range μm	Number of Particles (ml^{-1})
0.05-0.10	975
0.10-0.15	172
0.15-0.20	56
>0.20	36

5.5   Conclusions

A stable MFI-UF* value was obtained within 20-50 hours for polyacrylonitrile membranes. In contrast, the MFI-UF value for polysulphone membranes continuously decreased over time. Furthermore, polyacrylonitrile membranes were easier to clean, as one cleaning with sodium hypochlorite restored the clean water flux to ≈100%. Whereas, polysulphone membranes needed repeated chemical cleanings to restore the clean water flux.

As expected smaller particles were retained in the MFI-UF test, indicated by the significantly higher MFI-UF* values (2000-13 300 s/l^2) in comparison to the MFI$_{0.45}$ (1-5 s/l^2) for tap water. The MFI-UF* obtained appeared rather independent of MWCO as the MFI-UF* varied to only a small extent, from 2000 to 4500 s/L^2 for PAN and PS membranes ranging in MWCO from 3 to 100 kDa. Only the 1, 2, 5 kDa polysulphone membranes from manufacturer A gave markedly higher MFI-UF* values of 8400 - 13 300 s/L^2.

The polyacrylonitrile membrane surfaces were shown to be homogeneously permeable from FESEM micrographs. The porosity and pore size ranking determined from FESEM of 13 kDa < 6 kDa < 50 kDa did not agree with the increasing MFI-UF* value in the order PAN 6 kDa < 13 kDa ≈ 50 kDa. The low MFI-UF* value obtained with the 6 kDa membrane was attributed to its low flux and less cake compression.

FESEM micrographs of the 1, 2, 5 kDa polysulphone membranes from manufacturer A showed them to be heterogeneously porous due to pores limited to striated regions on the membrane surfaces. Consequently, artificially high MFI-UF* values were found for these membranes due to their low "average" surface porosity (lower filtration area) and not due to the retention of smaller particles. In contrast, no striations were found on the surfaces of the other polysulphone membranes.

Cake filtration was the dominant filtration mechanism for the polyacrylonitrile membranes, demonstrated by linearity in the t/V versus V plot. Whereas, for the polysulphone membranes, especially the 1, 2, 5 kDa membranes, cake filtration was difficult to demonstrate. This may be attributable to their low surface porosity and/or irregular distribution of pores over the membrane surface which caused a significant lateral flow component in the streamline flow and an increase in effective filtration area over time. Thus, the steadily decreasing MFI-UF value observed over time. Once these effects can be neglected, most likely the cake acts as a second membrane, determining the size of particles retained and hence the resultant MFI-UF* value obtained. This explains why the MFI-UF* appeared independent of MWCO for the 3 - 100 kDa membranes.

The PAN 13 kDa membrane was proposed as the most suitable reference membrane for application in the MFI-UF test. Cake filtration was proven to occur which gave a stable MFI-UF* value. Moreover, the pores of the PAN 13 kDa membrane are circa 1000 times smaller than the pores of the existing $MFI_{0.45}$ test membrane and thus will include smaller particles in the measurement.

Symbols

A	cross section area/membrane surface area [m^2]
A_o	reference surface area of a 0.45μm membrane filter [13.8×10^{-4} m^2]
b	gradient of line defined in Equation 5.7
C_b	concentration of particles in feedwater [kg/m^3]
d_p	particle diameter [m]
I	index for the propensity of particles in water to form a layer with hydraulic resistance [$1/m^2$]
J	linear fluid velocity/permeate water flux [m^3/m^2s]
K	Permeability constant in Equation 5.1
k	filtration constant in blocking laws
L	length of porous medium [m]
n	filtration constant in blocking laws [-]
ΔP	applied transmembrane pressure [bar] or [N/m^2]
ΔP_o	reference applied transmembrane pressure [2 bar]
Q	flow at temperature T [l/h]
R	resistance to filtration [1/m]
R_b	resistance due to blocking [1/m]
R_c	resistance of the cake [1/m]
R_m	membrane filter resistance [1/m]
t	filtration time [s]
T	temperature [°C]
V	filtrate volume [m^3]

Greek Symbols

α	specific cake resistance [m/kg]
ε	cake/membrane surface porosity [-]
η	fluid viscosity [Ns/m^2]
$\eta_{20°C}$	water viscosity at 20°C [Ns/m^2]
η_T	water viscosity at temperature (T) [Ns/m^2]
ρ_p	density of particles forming the cake [kg/m^3]
τ	tortuosity of membrane pores [-]
Δx	membrane thickness [m]

Abbreviations

CWF   clean water flux at 1 bar [l/m^2 h]
FESEM Field Emission Scanning Electron Microscope
$MFI_{0.45}$ Modified Fouling Index using a $0.45\mu m$ microfilter
MFI-UF Modified Fouling Index using an ultrafiltration membrane
SDI   Silt Density Index

References

1.   J. Granger, J. Dodds and D. Leclerc, Filtration of low concentrations of latex particles on membrane filters, Filt. Sep. (1985), 58-60.
2.   K.J. Kim, V. Chen and A.G. Fane, Ultrafiltration of colloidal silver particles: flux, rejection, and fouling, J. Coll. Inter. Sci., 155, (1993) 347-359.
3.   S.S. Madaeni, The ultrafiltration of very dilute colloidal gold suspensions, J. Porous Mat., 4, (1997), 31-44.
4.   J.C. Schippers and J. Verdouw, The modified fouling index, a method of determining the fouling characteristics of water, Desal., 32, (1980), 137-148.
5.   J.C. Schippers, J.H. Hanemaayer, C.A. Smolders and A. Kostense, Predicting flux decline of reverse osmosis membranes, Desal., 38, (1981), 339-348.
6.   F.P. Cuperus and C.A. Smolders, Characterization of UF membranes. Membrane characteristics and characterization techniques, Adv. in Coll. Inter. Sci, 34, (1991) 135-173.
7.   K.J. Kim, A.G. Fane, C.J.D Fell, T. Suzuki and M.R. Dickson, Quantitative microscopic study of surface characteristics of ultrafiltration membranes, J. Mem. Sci., 54, (1990) 89-102.
8.   M. Cheryan, Ultra filtration Handbook, (1986), Techonomic Publishing Company Co. Inc., Lancaster, USA, 1-375.
9.   M. Mulder, Basic Principles of Membranes Technology, Kluwer Academic Publisher, London (1990).
10.   A.G. Fane, C.J.D. Fell and A.G. Waters, The relationship between membrane surface pore characteristics and flux for ultrafiltration membranes, J of Mem. Sci., 9, (1981) 245-262.
11.   K.J. Kim, M.R. Dickson, V. Chen and A.G. Fane, Applications of field emission scanning electron microscopy to polymer membrane research, Micron Microscopica Acta 23, 3,(1992) 259-271.
12.   K.J. Kim, A.G. Fane, R. Ben, M.G. Liu, G. Jonsson, I.C. Tessaro, A. P. Broek and D. Bargeman, A comparative study of techniques used for porous membrane characterization: pore characterization, J. Mem. Sci., 87, (1994) 35-46.
13.   K.J. Kim and A.G. Fane, Low voltage scanning electron microscopy in membrane research, J. Mem. Sci.,88, (1994) 103-114.
14.   D.R. Sperry, The principles of filtration, Chem and Met. Eng., 15 (1916) 198-203.
15.   B.F. Ruth, G.H. Montillon and R. E. Montana, Studies in filtration, Ind. Eng. Chem., 25 (1933) 76.
16.   P.C. Carmen, Fluid flow through granular beds, Trans. Inst. Chem. Engrs. 15, (1937), 150-166.
17.   P.C. Carmen, Fundamental principles of industrial filtration, Trans. Inst. Chem. Engrs. 16, (1938), 168-188.
18.   A.J.V. Underwood, A critical review of published experiments on filtration, Trans. Inst. Chem. Eng., 4 (1926), 19.
19.   P.H. Hermans and H.L. Bredée, Principles of the mathematical treatment of constant pressure filtration, J. Soc. Chem. Ind., (1936) 1-4.
20.   V.E. Gonsalves, A critical investigation of the viscose filtration process, Recueil, 69 (1950) 873-903.
21.   J. Hermia, Étude analytique des lois de filtration à pression constante, Rev. Univ. Mines, 2 (1966) 45-51.
22.   J. Hermia, Constant pressure blocking filtration laws-application to power- law non-newtonian fluids, Trans. I. Chem Eng., 60 (1982) 183-187.
23.   P.M. Heertjes, Filtration, Trans. Instn. Chem. Engrs., 42 1964, T266-274.
24.   V. Lahoussine-Turcaud, M.R. Wiensner, J. Bottero and J. Mallevialle, Coagulation pretreatment for ultrafiltration of surface water, J. AWWA, December (1990), 67 –81.

25.   C.K. Lee and J. Hong, Characterization of electric charges in microporous membranes, J. Mem. Sci.,39, (1988) 79-88.

26.   M. Pontié, X. Chasseray, D. Lemordant and J.M. Lainé, The Streaming potential method for the characterization of ultrafiltration organic membranes and the control of cleaning treatments, J. Mem. Sci.,129, (1997) 125-133.

27.   J. Jacangelo and C. Buckley, Microfiltration, in water treatment membrane processes, J. Mallevialle, P.Odendaal, M.Wiesner [ed], American Water Works Association Research Foundation, Lyonnaise des Eaux, Water Research Commission of South Africa, McGraw-Hill, (1996).

28.   J.M. Lainé, J.P. Hagstrom, M.M. Clark and J. Mallevialle, Effects of ultrafiltration membrane composition, J. AWWA, 81 11 (1989).

29.   J.M. Lainé, M.M. Clark and J. Mallevialle, Ultrafiltration of lake water: effect of pretreatment on the partitioning of organics, THMP, and Flux, J. AWWA, December (1990) 82-87.

30.   A.S. Jonsson and B. Jonsson, The influence of nonionic and ionic surfactants on hydrophobic and hydrophilic ultrafiltration membranes, J. Mem. Sci., 56, (1991) 49-76.

31.   G. Crozes, C. Anselme and J. Mallevialle, Effect of adsorption of organic matter in fouling of ultrafiltration membranes, J. Mem. Sci., 84, (1993) 61-77.

32.   G.B. van der Berg and C.A. Smolders, Flux decline in ultrafiltration processes, Desal., 77, (1990) 101-133.

33.   J.A. Howell and M. Nystrom, Membranes in bioprocessing; theory and applications, J.A.Howell, V.Sanchez, R.W.Field [ed] Chapman and Hall, (1993).

34.   C.C. Ho and A.L. Zydney, Effect of membrane morphology on the initial rate of protein fouling during microfiltration, J. Mem. Sci., 155, (1999) 261-275.

35.   C.C. Ho and A.L. Zydney, Theoretical analysis of the effect of membrane morphology on fouling during microfiltration, Sep. Sci. tech., 34, 13 (1999) 2461-2483.

36.   A.G. Fane, Ultrafiltration: Factors influencing flux and rejection in R.J. Wakeman, [ed], Progress in Filtration and Separation, Vol. 4, Elsevier Science, Amsterdam, 101-179, (1986).

37.   N.M. Jackson, G.A. Davies and D.J. Bell, The Prediction of flux decline and blinding in cellular ceramic microfiltration membranes, Sep. Sci, Tech., 30 (7-9) (1995) 1529-1553.

38.   W. Pusch Membrane structure and its correlation with membrane permeability., J. Mem. Sci., 10, (1982) 325-360.

6

The MFI-UF as a Water Quality Test and Monitor

Chapter 6 is based on: "Monitoring Particulate Fouling in Membrane Systems" by Ś. F. E. Boerlage, M.D. Kennedy, M. p. Aniye, E. M. Abogrean, G. Galjaard, and J.C. Schippers. Published in Desalination, Vol. 118 pp. 131-142 (1998).

Contents

Abstract

In Chapter 5 the Modified Fouling Index using ultrafiltration membranes (MFI-UF) was developed to incorporate smaller particles into the MFI measurement. The polyacrylonitrile 13 kDa membrane (PAN 13 kDa) was proposed as the reference membrane for the MFI-UF test. This research investigates various aspects of the new MFI-UF test to establish its general use for characterising the fouling potential of feedwater; namely (i) proof of cake filtration via the stability of the MFI-UF over time and linearity of the index with particulate concentration, (ii) reproducibility of the MFI-UF with PAN 13 kDa membrane manufacture and membrane reuse (iii) the temperature and pressure dependency of the MFI-UF. The aforementioned aspects were examined using low (e.g. tap water) and high fouling feedwater (e.g. diluted canal water). Cake filtration was demonstrated for all feedwater tested. The MFI-UF was stable over time and proportional to particulate concentration. Reproducibility of the MFI-UF was found for 83% of the membranes tested and in five tests using one membrane, applying chemical cleaning in between. Correction to the reference temperature (20°C) of the MFI-UF test required only correction of the feedwater viscosity. However, cake compression with pressure was demonstrated for all the feedwater tested. Therefore, compressibility coefficients were determined for a given feedwater and a global compressibility coefficient was calculated for correction to the reference pressure (2 bar) of the MFI-UF test. Application of the MFI-UF as a continuous monitor of feedwater quality was trialed and requires further investigation.

6.1   Introduction

In the previous chapter the Modified Fouling Index using ultrafiltration membranes (MFI-UF) was developed to incorporate smaller particles present in a feedwater into the MFI measurement. A polyacrylonitrile 13 kDa membrane (PAN 13 kDa) was proposed as the most promising reference membrane for application in the newly developed MFI-UF test. The MFI-UF and the existing MFI are based on the assumption that cake filtration occurs for at least a part of the filtration time during measurement. Cake filtration was proven to occur after 20 hours for the PAN 13 kDa membrane by linearity in the t/V versus V plot which gave a stable MFI-UF over time for tap water filtration (refer section 5.4.1 in Chapter 5). However, only one feedwater was applied and other feedwater of a low, moderate and high fouling potential need to be investigated in order to verify that a stable MFI-UF value is also obtained with these feedwater. Moreover, a crucial requirement of the MFI-UF test, is that the MFI-UF index must be linear with the concentration of particles in a given feedwater (which is also an indirect proof of cake filtration). This was demonstrated for the existing MFI test and needs to be confirmed for the new MFI-UF test [1].

Another aspect which must be satisfied in order for the MFI-UF test to be operated as a water quality indicator is the reproducibility of the MFI-UF index using different PAN 13 kDa membranes and with repeated use of a single membrane. If the MFI-UF is sensitive to changes in membrane pore size or surface porosity, a different MFI-UF value will be obtained with different batches of PAN 13 kDa membranes for the same feedwater. In addition, as ultrafiltration membranes are expensive, for the test to be commercially viable the membranes must be reusable after cleaning and the measurement reproducible.

Furthermore, the process conditions e.g. temperature and pressure under which the MFI-UF is determined may have an effect on the resultant MFI-UF. The MFI-UF is carried out under constant pressure filtration, and the MFI-UF is corrected to standard reference conditions of 2 bar transmembrane pressure and a temperature of 20°C. Currently, the existing method to correct the MFI-UF value from the ambient feedwater temperature at which it was determined to the reference value of 20°C is based only on correcting the feedwater viscosity to 20°C. This accounts for a change in the rate of cake build up over time on the membrane while the cake itself is generally considered to be unchanged. However, temperature may have a direct effect on the specific cake resistance, measured in the MFI-UF value. A temperature decrease may cause the cake to shrink giving a denser cake with a higher specific resistance (and hence MFI-UF). Furthermore, a temperature increase may indirectly affect the specific cake resistance by causing expansion of the ultrafiltration membrane pores which lowers the resistance of the membrane to flow, but may allow smaller particles to pass the membrane. Hence, the size of particles retained may vary with temperature. Therefore, temperature correction of feedwater viscosity only, may not be sufficient in the MFI-UF test.

The MFI-UF test is ideally carried out at the standard reference transmembrane pressure of 2 bar. However, in some instances such as a very fouling water, it may be desirable to carry out the test at a lower test pressure. If the cake is composed of compressible particles then an increase in the applied transmembrane pressure causes the formation of a denser cake with a higher resistance

[2-9]. In previous research the existing MFI was found to depend on pressure and the extent of cake compression was determined using the compressibility coefficient, which relates the specific resistance of the cake to the applied pressure [7]. As in the case of temperature an increase in the applied pressure may affect not only the resistance of the cake but also the resistance of the ultrafiltration membrane [5,6,9-11]. Thus, the pressure dependency of both the membrane and the cake formed on the membrane needs to be ascertained and a method for pressure correction of the MFI-UF test developed.

This research aims at further developing the MFI-UF test to characterise the fouling potential of a feedwater employing the reference PAN 13 kDa membrane proposed in Chapter 5. Firstly, the occurrence of cake filtration will be verified for feedwater of varying fouling potential by examining the stability of the MFI-UF value over time and by determining if the MFI-UF index is linear with particle concentration for a given feedwater type. Secondly, the reproducibility of the MFI-UF with (i) membrane manufacture by determining the MFI-UF with different membranes from different batches employing the same feedwater and (ii) reuse of a single membrane by either backwashing or chemical cleaning of the membrane between measurements. Thirdly, the effect of temperature and pressure on the MFI-UF test will be examined, isolating their effect on both membrane resistance and the specific cake resistance. Methods will be developed to correct the MFI-UF value when the test is carried out at non reference conditions of temperature and pressure. If cake compression occurs with pressure a wide range of feedwater will be examined in order to estimate a global compressibility coefficient value, that can be incorporated in the pressure correction factor. Finally, the application of the MFI-UF test as a monitor to detect feedwater changes over time will be examined.

6.2   MFI Background

6.2.1   Cake Filtration

The Modified Fouling Index (MFI) is based on cake filtration theory, whereby, particles are retained on a membrane during filtration by a mechanism of surface deposition. The MFI is defined as the gradient of the linear region found in the plot of t/V versus V from the general cake filtration equation for constant pressure (refer Chapter 5 section 5.2.1) [1,4]:

$$\frac{t}{V} = \frac{\eta\,R_m}{\Delta P\,A} + \underbrace{\frac{\eta\,\alpha\,C_b}{2\Delta P\,A^2}}_{MFI}\,V \qquad\qquad ...(6.1)$$

Where V is filtrate volume, t the filtration time, α is the specific resistance of the cake deposited and C_b is the concentration of particles in the feedwater. The MFI is calculated at standard reference values of 2 bar transmembrane pressure (ΔP), a feedwater viscosity of $20\,^{\circ}C$ ($\eta_{20^{\circ}C}$) and the surface area of the $0.45\,\mu m$ microfiltration test membrane (A). R_m, assumed to be constant in the MFI, is the membrane resistance and is a function of properties such as thickness (Δx), surface porosity (ε), pore radius (r_p), and tortuosity (τ):

$$R_m = \frac{8 \Delta x \tau}{\varepsilon r_p^2} \qquad \qquad ...(6.2)$$

In the MFI, the fouling index (I) is defined as the product of α and C_b in the feedwater [1,4].

$$I = \alpha C_b \qquad \qquad ...(6.3)$$

The MFI assumes the retention of particles is constant and α has a time independent permeability and uniform cake porosity throughout the entire depth of the cake i.e. the cake is incompressible. However, when the cake is compressible the concept of the MFI is still valid as discussed later in section 6.2.2. The specific cake resistance can be related to particle properties and the porosity of the cake (ε) by the Carman-Kozeny relationship for spherical particles (Equation 6.4) where ρ_p is the density of particles forming the cake [9,10]. A decrease in cake porosity or a decrease in particle diameter size (d_p) leads to a significant increase in α.

$$\alpha = \frac{180 \, (1 - \varepsilon)}{\rho_p \, d_p^{\,2} \, \varepsilon^3} \qquad \qquad ...(6.4)$$

The t/V vs V plot, from which the MFI is determined, typically shows three regions in a MFI test using the 0.45μm membrane which correspond to (i) blocking filtration, (ii) cake filtration with or without compression and (iii) cake clogging and/or cake compression (refer Figure 6.1). The first sharp increase in slope is attributed to membrane pore blocking i.e. deposition of particles inside pores or blocking pore entry, followed by cake filtration which is a linear region of minimum slope. Alternatively, in a plot of MFI over time, cake filtration is observed as a minimum or stable MFI value depending on the length of cake filtration. The MFI test assumes that at least during a period of some significance ideal cake filtration takes place.

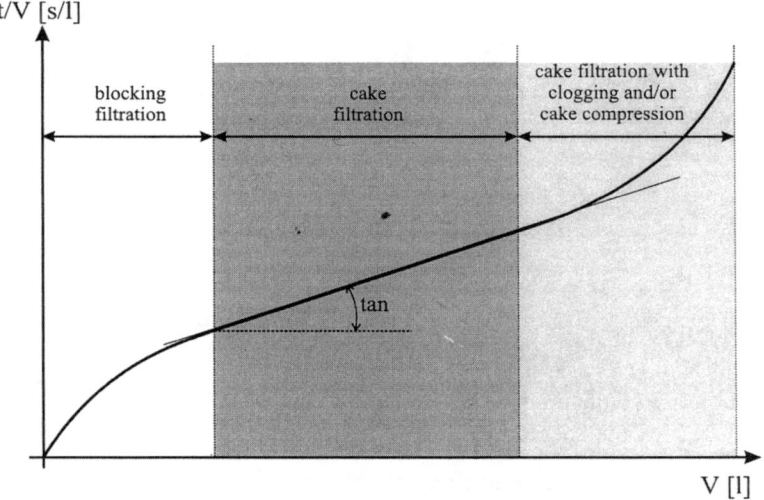

Figure 6.1:   Ratio of filtration time and filtrate volume as a function of total filtrate volume (based on a constant applied transmembrane pressure).

6.2.2   Cake Filtration with Compression

At the beginning of filtration the membrane resistance controls the rate of flow, and the whole of the pressure drop occurs in the membrane. As a cake builds up a proportion of the pressure drop is absorbed by the cake, and increasingly so as filtration proceeds, such that at a given point the membrane resistance becomes negligible. A hydraulic pressure gradient over the cake depth develops resulting in an increase in the viscous drag at the particle surfaces over which the fluid flows [3]. If the shape of the particles or their physical strength is such that the packing arrangement in the cake formed on the membrane can sustain this drag force without significant deformation, the cake is regarded as incompressible [3]. The porosity of the cake and its specific resistance are then independent of the imposed transmembrane pressure and Equation 6.1 can be applied.

However, in practice few cakes are incompressible. In membrane filtration many cakes are composed of clays and microbial cells which are highly compressible [7-9]. Thus, as the pressure drop increases over the cake during filtration (or for a test conducted at a higher applied pressure) the cake porosity reduces as particles compress creating a non uniform porosity distribution in the direction of flow [13]. Consequently, the specific cake resistance (and MFI) increases. In addition, the filtration of fines in the cake structure may also be responsible for an increase in the specific cake resistance over time [3,4,13,14]. These fines will block or narrow the voids present in the cake and is commonly referred to as cake clogging. Either of the two aforementioned effects i.e cake compression alone or in combination with cake clogging may lead to the third region in the t/V versus V plot in Figure 6.1.

It has also been suggested in literature that flux may cause cake compression. At a higher applied flux, Ruth [5] indicated that the initial void volume decreased resulting in a higher initial specific cake resistance. Alternatively, in constant pressure filtration "retarded packing compressibility" may occur [15].

Ruth [5] assumed negligible compression due to flux and expressed compressibility using the average specific cake resistance (α) for the pressure drop over the cake ΔP_c:

$$\alpha = \frac{\Delta P_c}{\displaystyle\int_0^{\Delta P_c} \frac{dP_c}{\alpha^1}} \qquad \qquad ...(6.5)$$

where α^1 is the local value of specific cake resistance. It is often assumed that the average specific resistance is constant (i.e. $\alpha = \alpha^1$) in order for Equation 6.1 to be applied. Alternatively, an empirical relationship (Equation 6.6) developed by Lewis [2] is widely applied to account for pressure which relates the specific resistance to pressure to the power of a compressibility coefficient (ω) and a constant (α_o). For *incompressible* cakes, ω is zero and the higher the compressibility coefficient the more compressible the cake [4,5].

$$\alpha = \alpha_o \, \Delta P_c^{\omega} \qquad \qquad ...(6.6)$$

When α is substituted into Equation 6.1 this results in:

$$\frac{t}{V} = \frac{\eta R_m}{\Delta P A} + \underbrace{\frac{\eta \alpha_o \Delta P_c^{\omega} C_b}{2 \Delta P A^2}}_{MFI} V \qquad ...(6.7)$$

as soon as ΔP_c is dominant a linear relationship between t/V versus V will appear and the MFI value can be determined. Previous research by Schippers [7] demonstrated that the cake formed on the 0.45μm microfiltration membranes was very compressible for colloidal solutions of formazine with a compressibility coefficient determined of 0.62 using Equation 6.6. Thus, the fouling index I can then be defined to incorporate the compressibility coefficient transforming Equation 6.3 into [7]:

$$I = \alpha_o \Delta P_c^{\omega} C_b \qquad ...(6.8)$$

6.3   Materials and Methods

6.3.1   Membranes

Seven polyacrylonitrile hollow fibre ultrafiltration membranes with a molecular-weight-cut-off of 13 000 Dalton (abbreviated as PAN 13 kDa) were used in the experiments (refer Table 6.1) from four different batches i.e. manufactured in the same month. Membranes designated with the same letter are from the same batch. The clean water flux (CWF) reported in Table 6.1 were measured in the MFI-UF test equipment (refer Figure 5.1 Chapter 5) using RO permeate and calculated according to Equation 6.9 corrected to 20°C and a transmembrane pressure of 1 bar.

$$CWF = \frac{\eta_T}{\eta_{20°C}} \frac{Q}{A \Delta P} \qquad ...(6.9)$$

where Q is the clean water flow at temperature T, and A is the membrane surface area which is 0.2 m^2 for the PAN 13 kDa membrane.

6.3.2   Feedwater

Feedwater tested included (i) conventionally pretreated River Rhine water (WRK-I) i.e. coagulation, sedimentation and filtration, used by Amsterdam Water Supply Company and WRK-I after further pretreatment steps in the following sequence; (ii) ozonation (iii) biological activated carbon filtration and (iv) slow sand filtration, (v) Delft tap water (vi) Delft canal water and at the Provincial Water Supply Company of North Holland: (vii) IJssel Lake water which has undergone conventional pretreatment plus an additional rapid sand filtration polishing step (WRK-III) and (viii) membrane concentrate from the reverse osmosis installation. Only canal water was pretreated prior to the MFI-UF measurement, by filtering through a 60μm nylon mesh to remove large contaminants.

Table 6.1:   Clean Water Flux (CWF) of PAN 13 kDa ultrafiltration membranes.

Membrane Code:	Clean Water Flux at 20°C and 1 bar [l/m² h]
A1	174
A2	165
A3·	165
B1	130
C1	165
C2	165
E1	152

6.3.3   MFI-UF Determination

For the MFI-UF test the feedwater was pumped to the UF membrane inlet at a constant applied transmembrane pressure (ΔP 0.5 - 2.0 bar) using the MFI-UF equipment as described in Chapter 5. The MFI-UF was calculated by a software programme according to Equation 6.10.

$$MFI-UF \; = \; \underset{\uparrow}{\frac{\eta_{20°C}}{\eta_T}} \; \underset{\uparrow}{\left(\frac{\Delta P}{\Delta P_o}\right)^{1-\omega}} \; \underset{\uparrow}{\left(\frac{A}{A_o}\right)^2} \; \frac{d\frac{t}{V}}{dV}$$

$$\text{T Corr} \qquad \text{P Corr} \qquad \text{A Corr} \qquad\qquad ...(6.10)$$

which includes correction factors (Corr) to standard reference conditions. Where the temperature correction factor (T Corr) corrects the feedwater viscosity to 20°C, the pressure correction factor (P Corr) corrects the applied transmembrane pressure to 2 bar (ΔP_o) and ω refers to the compressibility coefficient which can be used to correct for cake compression. The area correction factor (A Corr) corrects the ultrafiltration membrane area to the reference surface area of the $MFI_{0.45}$ microfiltration membrane ($A_o = 13.8\times10^{-4}$ m²). Thus, the $MFI_{0.45}$ and MFI-UF are directly comparable. The MFI-UF value reported represents the mean value calculated from the stable region of the MFI-UF over time in a single test.

6.3.4   Membrane Cleaning

After a MFI-UF test the membranes were backwashed (1.5 bar) with RO permeate (ambient temperature) for 15 minutes. In the case of chemical cleaning, backwashing was followed by a chemical cleaning using a 200 ppm sodium hypochlorite solution which was recirculated for 1 hour at 1 bar. Sodium hypochlorite was prepared from analytical grade reagent and RO permeate. To remove any chemicals the membranes were backwashed again with RO permeate before the CWF restoration was measured.

6.3.5 Linearity Experiments

In the linearity experiments one membrane (A2) was employed and after a test the CWF was restored to 100% before subsequent testing. Linearity experiments were carried out with the MFI-UF determined for the most concentrated feedwater first, after which the feedwater was serially diluted and tested. The other feedwater; tap water, WRK-I and canal water, were collected and allowed to equilibrate to ambient temperature. The diluents, RO permeate (for WRK-I and tap water) and tap water (for canal water), were added to the calculated height of the 1m³ test tank for the desired dilution, stirred and allowed to temperature equilibrate. Serial dilutions were prepared taking into account the previous dilution. Whereas, the membrane concentrate was diluted on line by combining the flows of RO permeate with the membrane concentrate. Experiments for all feedwater were carried out at an applied transmembrane pressure of 1 bar. In the case of tap water, linearity experiments were also carried out at an applied transmembrane pressure of 0.5 and 1.6 bar.

6.3.6 Reproducibility Experiments

6.3.6.1 MFI-UF Reproducibility using Different Membranes

For the reproducibility of the MFI-UF with membrane manufacture, tap water was collected in a 1m³ tank and allowed to temperature equilibrate. During testing the tank contents were continuously stirred. Six membranes with 100% CWF were tested, three manufactured from one batch (A), three from other batches (B and C) using the collected tap water at an applied transmembrane pressure of 1 bar.

6.3.6.2 MFI-UF Reproducibility with Membrane Reuse

The reproducibility of the MFI-UF with membrane reuse was investigated using one PAN 13 kDa membrane in two series of experiments. In one series, the membrane was cleaned by backwashing (as described in section 6.3.4) between each measurement of WRK-I water which was applied on line. In the second series, membranes were chemically cleaned in addition to backwashing to give 100% CWF between measurements. In this case tap water was applied as feedwater which was collected and treated in the same way as described in 6.3.6.1.

6.3.7 Temperature and Pressure Experiments

6.3.7.1 Determining Temperature Dependency

The *membrane resistance* (R_m) of the A1 and B1 membranes (100% CWF) was determined using RO permeate at selected temperatures ranging between 17-36°C. The RO permeate was collected in a 200L tank and raised to the desired temperature (±0.5°C) using an electric coil with the excess flow recirculated back to the tank. Between each measurement the membrane was cleaned to 100% CWF restoration. The flux (J) through the membrane was measured for one hour at a transmembrane pressure (ΔP) of 1 bar and calculated by the software programme to a reference value of 20°C. R_m was calculated from Darcys Law according to Equation 6.11 (Chapter 5.2.1) assuming no fouling and thus no blocking or cake resistance and the viscosity of the feedwater η_T at temperature T.

$$R_m = \frac{\Delta P}{\eta_T \times J} \qquad\qquad ...(6.11)$$

The temperature dependency of the *MFI-UF* was examined by tap water filtration at selected temperatures ranging between 15–31 °C. Tap water was collected in a 200L tank, with the temperature controlled using an electrical coil and maintained at ±0.5 °C by the addition of hot tap water. The MFI-UF value was measured until a stable MFI-UF value was obtained (8-10 hours) under a ΔP of 0.75 bar and calculated by the software programme with and without the T Corr factor in Equation 6.10 ($\omega = 0$). The membrane was cleaned to 100% CWF restoration between each measurement.

6.3.7.2   Determining Pressure Dependency

The *membrane resistance* (R_m) of the A1 and E1 membranes (100% CWF) was determined using RO permeate at selected applied transmembrane pressures ranging between 0.5-2.0 bar. One membrane had been previously used with 100% CWF and the other was new. Membrane flux was measured at the desired ΔP until a stable flux was recorded, starting at the lowest ΔP. The ΔP was then increased to the next test pressure and continued until a stable flux was observed. This procedure was repeated to the maximum ΔP of 2 bar i.e. the reference pressure of the MFI-UF test. Membrane resistance at the test ΔP was calculated using Equation 6.11 with flux J corrected to 20 °C, and η_T is for 20 °C.

The pressure dependency of the *fouling index I* (and hence the MFI-UF value) was examined for a range of feedwater. All feedwater were applied on line with the exception of the diluted canal water (10% canal water with 90% tap water) which was treated in the same way as described in section 6.3.5. One membrane was used per feedwater and the MFI-UF value determined at selected ΔP ranging between 0.5-2.0 bar. Between measurements the membrane was chemically cleaned. The fouling index I was calculated by manipulation of Equation 6.10 from the MFI-UF value determined by the software. The I value reported is corrected to 20 °C and a pressure of 2 bar. To incorporate cake compression into the pressure correction factor, the compressibility coefficient (ω) was calculated from the log plot of the fouling index I for a feedwater versus the log applied pressure.

6.3.8   MFI-UF Water Quality Monitor Experiment

Initially tap water was fed on line to the MFI-UF membrane after which the feedwater was changed on line to a 15% canal water dilution. The diluted canal water was prepared as described in section 6.3.5. The feedwater was then changed from the diluted canal water back to tap water after *circa* 65 hours of filtration.

6.4   Results and Discussion

6.4.1   Cake Filtration

6.4.1.1   Stable MFI-UF Value over Time

The existing MFI and the MFI-UF under development are based on the assumption that cake filtration (with or without compression) occurs for at least a part of the filtration time during measurement. Results of the MFI-UF over time for various feedwater of different fouling potential are presented in Figure 6.2. Cake filtration was observed to occur when the MFI-UF value stabilised at 2, 4, and 10 hours for WRK-I water, tap water and 25% canal water, respectively. The MFI-UF value remained stable for the duration of the test, which in the case of WRK-I water was up to 60 hours of filtration (not shown). Therefore, cake filtration occurred for a prolonged period of time.

The most fouling feedwater (25% canal water) gave the highest MFI-UF value of 16 400 s/l^2. Whereas, the MFI-UF obtained for the less fouling tap and WRK-I water, were much lower at 4200 s/l^2 and 1860 s/l^2, respectively. The WRK-I water appears to be of a better quality than the tap water as measured by the MFI-UF test. The $MFI_{0.45}$ commonly measured for conventionally treated tap water is in the range of 1-5 s/l^2. The MFI-UF value is much higher for tap and WRK-I water due to the retention of smaller particles as discussed in Chapter 5 and can be considered as a more sensitive measure of the fouling potential of feedwater.

A typical example of the $MFI_{0.45}$ over time is presented in Figure 6.3 for WRK-III water (also conventionally pretreated). The real $MFI_{0.45}$ is defined as the minimum slope in the t/V versus V plot which corresponds to cake filtration. In the $MFI_{0.45}$ over time plot this is observed as a minimum. (The high $MFI_{0.45}$ value of 33 s/l^2 is due to problems occurring with the pretreatment scheme at that time as in general the $MFI_{0.45}$ of WRK-III water is between 3-5 s/l^2). Characteristically the minimum occurs shortly after commencing filtration and for only a short time in the $MFI_{0.45}$ versus time plot. In the case of WRK-III water the $MFI_{0.45}$ could be determined after 5 minutes of filtration and stabilised for only 10 minutes. Whereas, the MFI-UF took 2-10 hours of filtration before it could be determined and the period for which it was stable was dramatically longer. This behaviour can be explained by the higher flux of the microfiltration membrane $\approx 10\ 000\ l/m^2$ h bar which leads to faster cake development on the membrane surface. The rapidly increasing $MFI_{0.45}$ value after the stable period is most likely due to clogging of the cake as fines become entrapped in the cake itself and/or cake compression. No increase was observed in the MFI-UF value after stability had been reached for any of the feedwater tested. This is most likely a consequence of the lower flux of the UF membrane. It is expected that if filtration had been continued for a longer period of time clogging of the cake and/or cake compression might also have occurred.

The above results show that the MFI-UF can be applied as a relatively short test (2-10 hours) to characterise the fouling potential of a given feedwater. In addition, due to the prolonged stability of the MFI-UF value, the MFI-UF test could potentially be employed as a monitor to detect changes in the particulate fouling potential of a feedwater over time.

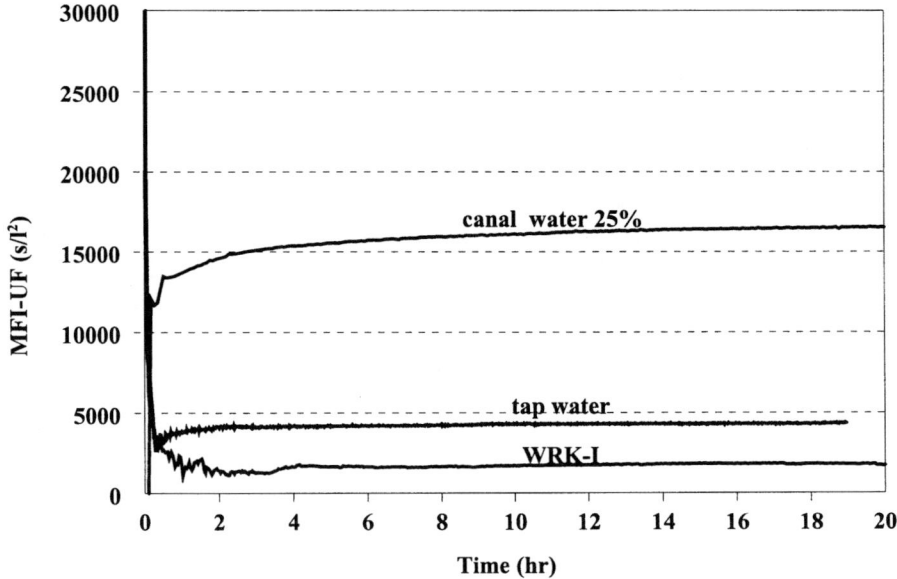

Figure 6.2:   MFI-UF measured over time for WRK-I water, tap water and a 25% canal water dilution.

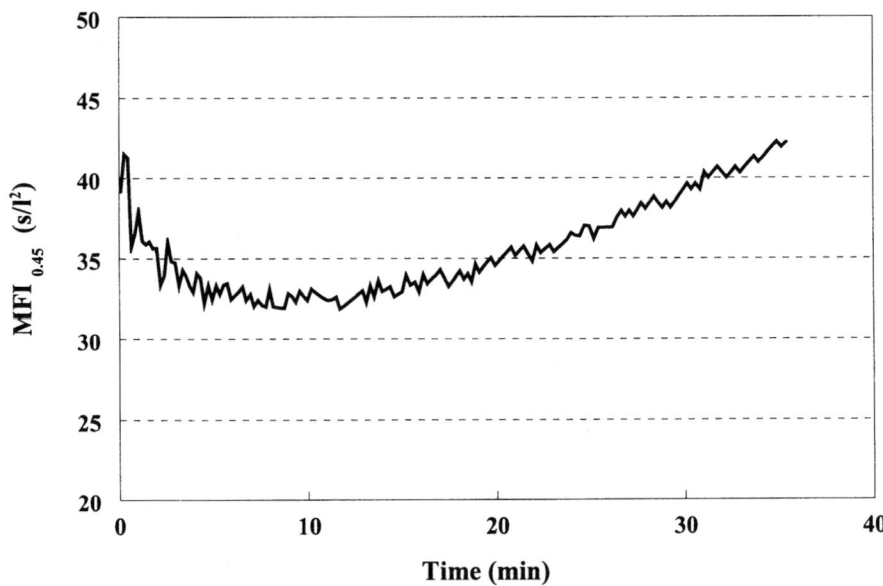

Figure 6.3: $MFI_{0.45}$ measured over time for WRK-III water.

6.4.1.2   Linearity of MFI-UF with Colloidal Concentration

Further evidence that cake filtration occurs during the MFI-UF test can be observed in the results of the MFI-UF index as a function of particulate concentration in the feedwater. This premise is based on the fouling index I being directly related to the concentration of particles C_b (Equation 6.3). Thus, I will decrease directly in proportion to an increase in the dilution factor of C_b while the specific cake resistance component (α), characteristic of a feedwater type and independent of concentration, remains constant. In Figure 6.4 the results of the MFI-UF with dilutions of membrane concentrate, WRK-I, tap and canal water conducted at an applied transmembrane pressure of 1.0 bar are shown. Linearity was found for all feedwater, the regression coefficient calculated for membrane concentrate, WRK-I, tap and canal water were 0.998, 0.993, 0.998 and 0.998, respectively. The MFI-UF for WRK-I water and the membrane concentrate was found to be directly related to the dilution factor $e.g.$ the MFI-UF value for WRK-I at 25% dilution is 590 s/L^2 and is approximately double at 50%, 1230 s/L^2. In contrast, the MFI-UF of tap water and canal water were higher than expected at lower dilution, however this maybe due to errors in dilution.

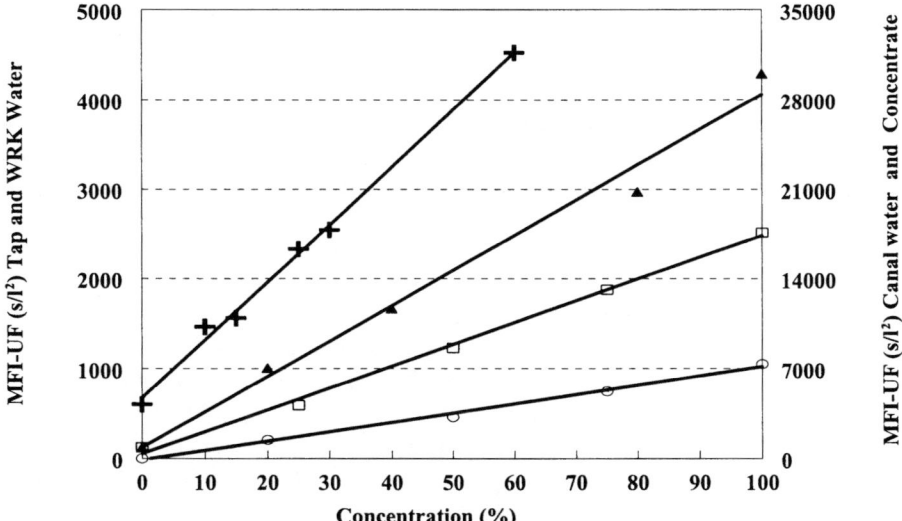

Figure 6.4:   Relationship of the MFI-UF value $vs.$ dilutions of feedwater; membrane concentrate (□), WRK-I (○), tap (▲) and canal water (✚).

A further test was carried out with dilutions of tap water to investigate if linearity was found at different pressures. Linear regression coefficients calculated for the MFI-UF values of the tap water dilutions at each applied transmembrane pressure were all greater than 0.995 (Table 6.2). Thus, the MFI-UF was found to be linear with particulate concentration for all feedwater tested and for tap water at applied transmembrane pressures within the range of 0.5-1.6 bar. These results further support cake filtration as the filtration mechanism operative in the MFI-UF test. Since the MFI-UF values in Table 6.2, are normalised to 2 bar, the higher values at higher applied transmembrane pressures indicate that cake compression occurs.

Table 6.2:   MFI-UF *vs.* dilutions of tap water (with RO permeate) at varying applied transmembrane pressure (0.5-1.6 bar ΔP) with linear regression coefficient (R^2).

% diluted Tap Water	MFI-UF s/L² at ΔP (bar)		
	0.5 ΔP	1.0 ΔP	1.6 ΔP
25	700	970	1600
50	1400	2680	3000
75	2100	4080	4770
100	2840	5320	6170
R^2	0.999	0.996	0.999

6.4.2   MFI-UF Reproducibility using Different Membranes

The mean MFI-UF value and 95% confidence interval calculated from MFI-UF measurements of tap water using six different PAN 13 kDa membranes, to test the reproducibility of the MFI-UF value, was 2970 ± 180 s/L². Differences in the clean water flux of the membranes were found, the maximum difference was 25% between A1 and B1 (refer Table 6.1). This may be attributable to differences in surface porosity and/or pore size and hence the membrane resistance R_m (Equation 6.2). However, no trend was observed in the MFI-UF related to the membrane clean water flux. Similarly, no clear trend was discernable for the MFI-UF measured for membranes within a batch and between batches (refer Figure 6.5). For example, the C1 membrane shows a markedly lower MFI-UF than the C2 membrane from the same batch, *circa* 10% difference. However, five of the six membranes fall within the 95% confidence interval calculated for the mean MFI-UF for the membranes. This indicates that manufacture of the PAN 13 kDa membrane was sufficiently reproducible to lead to a reproducible MFI-UF for a feedwater. Although, only one feedwater type was used to test the reproducibility of different membranes, it is expected that a similar MFI-UF would be obtained with other feedwater. However, further testing is recommended to confirm this.

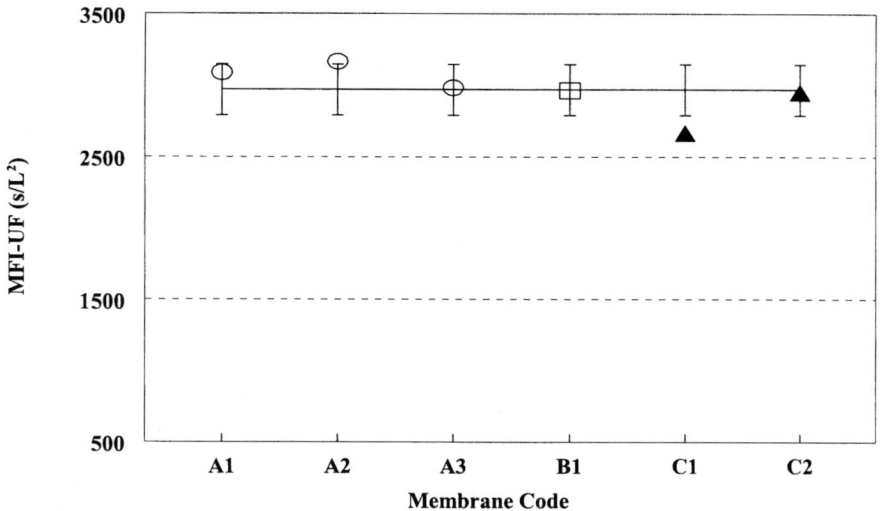

Figure 6.5: Reproducibility of the MFI-UF determined for tap water with membrane manufacture. Membrane batches A (O) B (□) and C (▲). Mean and 95% confidence interval indicated.

6.4.3 MFI-UF Reproducibility with Membrane Reuse

6.4.3.1 MFI-UF Reproducibility with Backwashing

The MFI-UF determined for WRK-I water with repeated testing using one PAN 13 kDa membrane is presented in Figure 6.6. The membrane was backwashed between measurements and the clean water flux restoration is also presented in Figure 6.6. The MFI-UF ranged between 1750-3650 s/l^2 with a mean MFI-UF of 2640 s/l^2 over 33 measurements. Only 10 of the measurements were found to fall within the 95% confidence interval ($\pm140s/l^2$) of the mean value. In general, the MFI-UF increased over the first 16 measurements which was accompanied by a loss in clean water flux restoration, indicating that pores may have become partially or fully blocked. The net effect of which is an increase in the MFI-UF in successive tests due to a loss in filtration area (Equation 6.1) and possibly due to the retention of smaller particles due to a decrease in the effective MWCO.

It was expected that after a number of tests the membrane would become "conditioned" as a stable effective MWCO was established, resulting in reproducibility of the MFI-UF. However, the reproducibility was poor and only 11 of the remaining 18 measurements fell within the mean and 95% confidence interval (2880 ± 140 s/l^2) calculated from the 16th to 23rd measurements. Moreover, the MFI-UF continued to fluctuate reaching a maximum value of 3650 s/l^2 in the 23rd measurement.

The MFI-UF appeared to be generally correlated with the clean water flux (CWF) restoration whereby the lower the CWF restoration, due to the low efficiency of backwash, the higher the MFI-UF obtained e.g. measurement 23. This sensitivity to the CWF restoration was not expected as in section 6.4.2, a reproducible MFI-UF was obtained using membranes A1 and B1, despite a 25% difference in their initial specific clean water flux. One explanation may be that if a membrane is not totally cleaned membrane pores may be blocked to a differing extent which affects the particles subsequently retained in the cake. Whereas, for clean membranes blocking may occur to a similar extent despite differences in their CWF which leads to a similar cake build up for the same water type as found in section 6.4.2.

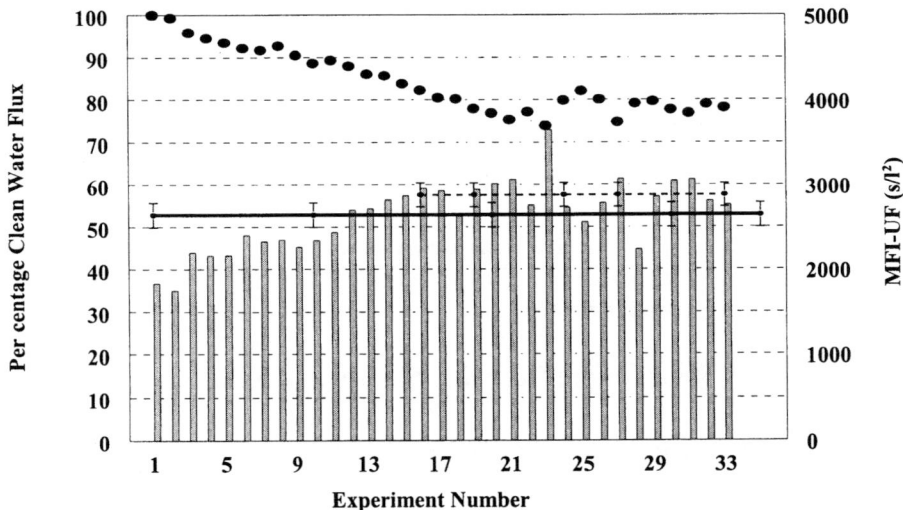

Figure 6.6:   MFI-UF (❙) determined in repeated measurements of WRK-I water using a single PAN 13 kDa membrane with backwashing in between and the per centage clean water flux restoration (●). Mean MFI-UF and 95% confidence interval indicated of measurements 1-33 (-) and 16-33 (---).

6.4.3.2   MFI-UF Reproducibility with Chemical Cleaning

In the previous section poor MFI-UF reproducibility was found with membrane reuse applying only simple backwashing between measurements. Therefore, chemical cleaning to 100% clean water flux restoration was employed between measurements. The MFI-UF obtained in five repeated measurements for tap water are presented in Figure 6.7. The MFI-UF was found to be reproducible, as all values fell within the 95% confidence interval of the mean MFI-UF (2640 ± 140 s/l^2) calculated for the five tests. Thus, chemical cleaning of the MFI-UF test membrane is recommended to allow membrane reuse. Alternatively, the development of cheap disposable PAN 13 kDa membranes for use in the MFI-UF test would be another option.

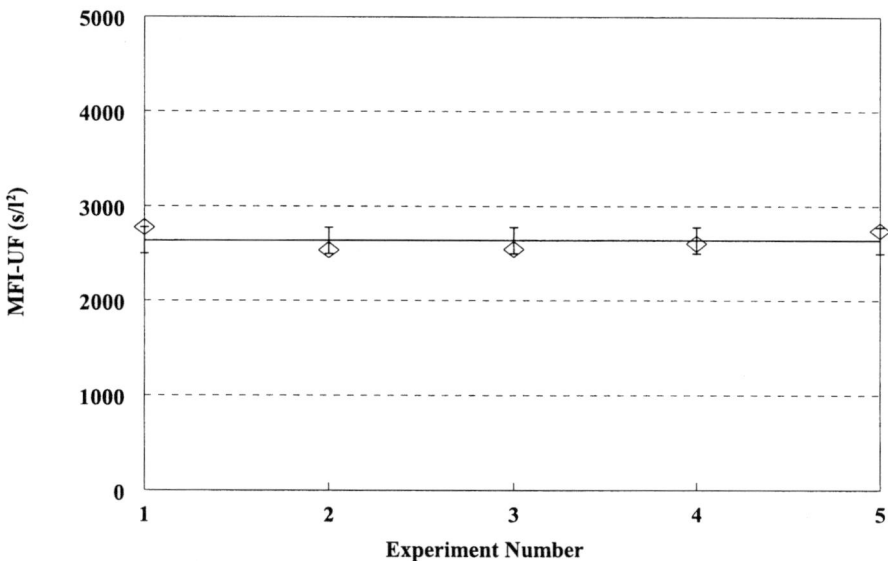

Figure 6.7: MFI-UF determined in repeated measurements of tap water using a single PAN 13 kDa membrane with chemical cleaning to 100% CWF in between. Mean MFI-UF and 95% confidence interval indicated.

6.4.4   Effect of Process Conditions on the MFI-UF

6.4.4.1   Temperature Dependency and Correction

The effect of temperature on the permeability of two PAN 13 kDa membranes (A1 and B1) expressed as membrane resistance (R_m) in the 17-36°C range is given in Table 6.3. No thermal expansion of the membrane pores was evident for an increase of up to 18°C in this temperature range as the A1 and B1 membrane resistance were found to be constant (after correction for the feedwater viscosity decrease (Equation 6.9) to the reference temperature of 20°C). A maximum increase of 0.7% was found for membrane B1, however, this is within experimental error. This result agrees with a study of Huisman [10] where no change in membrane resistance was found for polymeric ultrafiltration membranes in the temperature region tested of 4-20°C. Consequently, no effect is expected on the size of particles retained in the MFI-UF test due to changes in the membrane with increasing temperature.

The effect of a temperature increase of ≈16°C on the MFI-UF determined for tap water with and without viscosity correction of the feedwater to 20°C is given in Table 6.4. At a higher temperature the cake builds up at a quicker rate over time and without correction for the viscosity decrease in the feedwater flow, the MFI-UF decreases. Therefore, the feedwater appears less fouling at higher temperatures. After compensating for the differences in viscosity of the feedwater, the MFI-UF should be comparable if the specific resistance measured in the fouling index I is temperature independent. A maximum difference of 5% was observed for a temperature increase from 15 to 31°C which is within experimental error. Thus, temperature appears to have a negligible effect on the visco-elastic behaviour of cake particles and the packing (porosity) of the cake. Hence, the specific cake resistance and MFI-UF was temperature independent for the feedwater tested.

Table 6.3:   Membrane Resistance (R_m) as a function of temperature for two PAN 13 kDa membranes.

Membrane:	A1	B1	
Temperature (°C)	Membrane Resistance R_m (m^{-1}) × 10^{12}	Temperature (°C)	Membrane Resistance R_m (m^{-1}) × 10^{12}
17	2.01	19	2.75
24	2.02	26	2.76
30	2.02	31	2.76
35	2.01	36	2.77

Table 6.4:   MFI-UF determined for tap water calculated at ambient feedwater temperature (T) and corrected to a reference temperature of 20°C by correcting the viscosity of the feedwater.

Temperature T (°C)	MFI-UF at temperature T with no feedwater viscosity correction (s/l^2)	MFI-UF with feedwater viscosity correction to 20°C (s/l^2)
15	3060	2710
19	2820	2760
21	2740	2810
25	2470	2780
31	2230	2850

From the results presented in this section no temperature correction needs to be applied for membrane resistance or the specific cake resistance in the MFI-UF index for varying feedwater temperature. Only temperature correction for feedwater viscosity to the reference temperature of 20°C is required in the MFI-UF test.

6.4.4.2   Pressure Dependency and Correction

The effect of an increase in applied transmembrane pressure on the resistance of a used membrane with 100% clean water flux restoration (A1) and a new membrane (E1) is given in Table 6.5. The A1 and E1 membrane resistance both increased with pressure suggesting that the membrane compacted under the increased applied transmembrane pressure. Persson et al [11] fitted the pressure dependence of membrane resistance data for polymeric ultrafiltration membranes to a power law relationship for the pressure range of 0-3 bar. Whereas, Chellam et al [9] found a linear relationship of membrane resistance with pressure in the lower pressure range tested of 0-0.8 bar. Compaction of the PAN 13 kDa membranes was observed to follow a power law relationship as reported by Persson et al [11] as follows:

$$R_m = R_{mo} \Delta P^h \qquad\qquad ...(6.12)$$

where R_{mo} is the pressure- invariant component of the membrane resistance and is the membrane resistance at an applied transmembrane pressure of zero and h is the membrane compaction coefficient. The membrane compaction coefficient determined for the new membrane was 0.058. Most likely the compaction is reversible as the membrane compaction coefficient for the membrane which had been used prior to the compaction experiment in a number of MFI-UF determinations was similar (h = 0.052). In the study of Persson et al [11] a higher compaction coefficient, h = 0.8, was found, indicating significant compaction. Whereas, in this study the membrane compaction coefficient determined for the two polyacrylonitrile membranes was small. The initial membrane resistance increased by only 8% and 7% for the new and used membrane, respectively. This increase is not expected to have a significant effect on membrane surface properties such as pore size. Therefore, the size of particles retained in the MFI-UF test are expected to be constant for the 0-2 bar range and no effect on the MFI-UF value is expected due to membrane compaction.

Table 6.5:   Membrane Resistance (R_m) as a function of pressure of two PAN 13 kDa membranes; A1 previously used with 100% clean water flux restoration and E1 a new membrane.

Membrane:	A1		E1
Pressure (bar)	Membrane Resistance R_m (m^{-1}) × 10^{12}	Pressure (bar)	Membrane Resistance R_m (m^{-1}) × 10^{12}
0.5	1.93	0.5	2.26
0.7	1.97	-	-
1.1	1.98	1.0	2.33
1.6	2.05	1.5	2.40
2.0	2.08	2.0	2.45

A low fouling feedwater, tap water, was employed as a preliminary check of the pressure dependency of the specific cake resistance and hence the fouling index I in the MFI-UF test. In Figure 6.8 the fouling index I measured at various applied pressures in the range 0.5-2.0 bar is given. I includes a simple ratio correction factor for pressure (Equation 6.10) to a reference pressure of 2 bar. C_b can assumed to be constant for all the tests and the same membrane was employed to prevent any experimental artifacts occurring during the test. Therefore, in the absence of cake compression, the same value for I should have been found irrespective of the applied operating pressure. However, despite correction for pressure when the ΔP is increased from 0.5 to 1.0 bar, I is *circa* 2.1 times higher. Treated surface water in the Netherlands may contain a relatively large amount of considerably hydrated colloids [1]. Thus, at higher applied pressures these colloids, compress forming a denser cake due to a reduction in cake porosity, resulting in an increase in the specific cake resistance and the fouling index I. However, as previously mentioned an increase in flux may also cause cake compression [4,5]. Therefore, the structural arrangement and the extent of cake compression could be the combined result of an increase in the ΔP or the increased flux to the membrane as a result of the increased pressure.

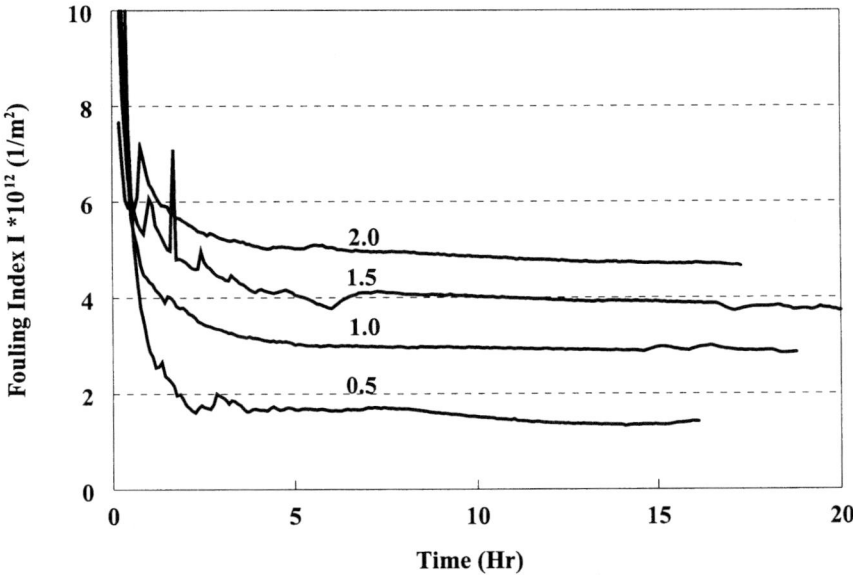

Figure 6.8:   The fouling index I measured over time for tap water at varying applied transmembrane pressure (ΔP 0.5-2.0 bar).

The pressure dependency of the specific cake resistance (measured in the fouling index I) for tap water was found to fit the empirical power function often quoted in literature for specific cake resistance (Equation 6.6) which is similar to the relationship for membrane resistance. The compressibility coefficient determined for the cake formed by the filtration of tap water was 0.82 (refer Figure 6.9) which suggests that the cake is very compressible.

Pressure correction factor assuming (i) the cake is incompressible ($\omega = 0$) and (ii) incorporating the compressibility coefficient ($\omega = 0.82$) were used to correct the experimentally determined MFI-UF of tap water at 0.5 and 1.0 bar to the reference pressure of 2.0 bar. A comparison of the two correction approaches, to the MFI-UF value experimentally determined at 2.0 is given in Table 6.6. When compression effects are ignored the corrected MFI-UF value was underestimated by $\approx 70\%$ and $\approx 38\%$ at 0.5 and 1.0 bar, respectively. Whereas, closer agreement was found when the compressibility coefficient was incorporated, the corrected MFI-UF value was underestimated by 10% and overestimated by 5% at 0.5 and 1.0 bar, respectively. Thus, compression needs to be accounted for in the pressure correction factor.

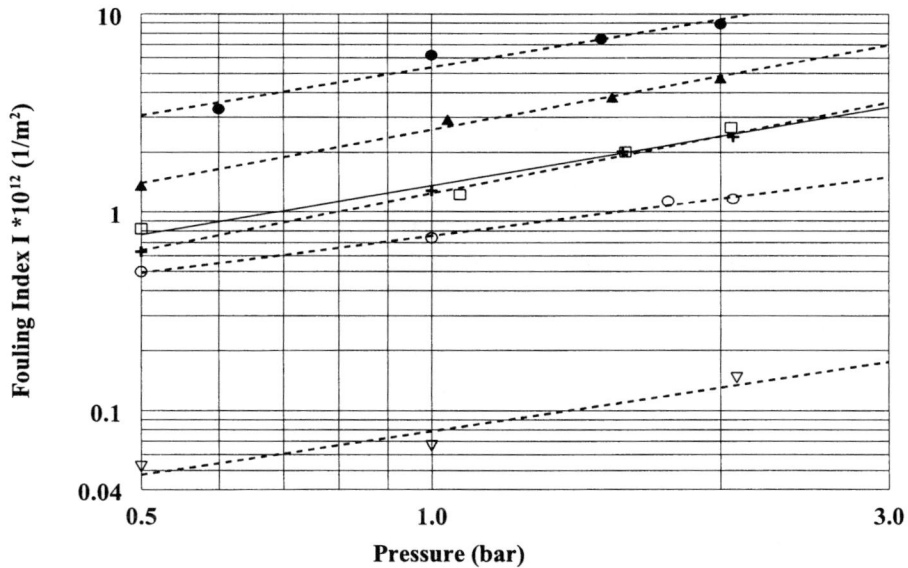

Figure 6.9: Log plot of fouling index I *vs* applied transmembrane pressure (0.5-2.1 bar) in the MFI-UF test for diluted canal water (10%) (●), tap water (▲), WRK-I water (✚) and WRK-I water after increasing levels of pretreatment; ozonation (□), biological carbon filtration (○) and slow sand filtration(▽).

Table 6.6: Pressure correction of the MFI-UF measured for tap water at applied transmembrane pressure in the range of 0.5 to 2.0 bar assuming that (i) cakes are incompressible ($\omega=0$) and (ii) incorporating the compressibility coefficient of $\omega=0.82$ into Equation 6.8

Pressure (bar)	MFI-UF (s/l²) no compression correction ($\omega = 0$)	MFI-UF (s/l²) compression correction ($\omega = 0.82$)
0.50	1800	5610
1.00	3860	6600
1.50	5000	6200
2.00	6260	6260

The compressibility coefficient is expected to vary from one water type to another. Therefore, in order to find a global compressibility coefficient for general use in the pressure correction factor, the compressibility coefficient was determined for a range of feedwater types. The dependence of the fouling index I on applied transmembrane pressure for the feedwater examined, including diluted canal water (10%), and WRK-I water from Amsterdam Water

Supply with increasing pretreatment levels, are presented in Figure 6.9. The compressibility coefficient obtained from Figure 6.9 for all feedwater are summarised in Table 6.7 and ranged from 0.51 to 0.85. In general, the compressibility coefficient was higher for feedwater with no pretreatment or lower levels of pretreatment i.e the diluted canal water, tap and WRK-I water (ω = 0.82-0.85). These feedwater are expected to contain particles of a wider size distribution and most likely contain a higher proportion of bigger particles which can easily deform under pressure. This will give an accompanying decrease in cake porosity and an increase in the fouling index I through an increase in the specific cake resistance. Whereas, ozonation will change the nature of chemical bonds in organic molecules and split larger particles into smaller particles which may be more resistant to compression. In addition to size, the nature of the particles themselves will also influence the observed compressibility coefficients. For instance bacterial debris, some types of organic matter and ferric hydroxide flocs are considered to be more compressible than just clay particles. Thus, the lowest compressibility factor which was found, that after biological activated carbon filtration, may be due to the removal of compressible organic matter through adsorption and biodegradation in the biological activated carbon filter.

Table 6.7:   The compressibility coefficient (ω) and linear regression coefficient (R^2) determined for various feedwater; diluted canal water (10%), tap water and WRK-I water after increasing levels of pretreatment; ozonation, biological activated carbon filtration (BACF) and slow sand filtration (SSF).

Feedwater	Compression Factor(ω)	Linear Regression Coefficient (R^2)
Diluted canal water	0.85	0.929
Tap water	0.82	0.981
WRK-I	0.88	0.990
After Ozonation	0.71	0.975
After BACF	0.51	0.982
After SSF	0.68	0.970

An average compressibility coefficient (0.75) was determined from the values recorded in Table 6.7, which was used as the global compressibility coefficient for pressure correction. The percentage difference of the corrected MFI-UF applying the global compressibility coefficient, the individual compressibility coefficient determined for a feedwater and for the base case of ω = 0 i.e. no cake compression to that experimentally determined at 2 bar is given in Table 6.8.

Table 6.8:   Difference (%) between the MFI-UF experimentally determined at 2 bar and the MFI-UF determined at applied transmembrane pressures in the range of 0.5-2.0 bar and corrected to 2 bar assuming (i) the cake is incompressible ($\omega = 0$) and correcting for cake compression by incorporating (ii) the global compressibility coefficient of $\omega = 0.75$ and (iii) the individual compressibility coefficient for a feedwater into Equation 6.10.

Feedwater	Applied Transmembrane Pressure ΔP	% difference from MFI-UF experimentally determined at 2 bar		
		no compression correction ($\omega = 0$)	global compressibility coefficient ($\omega = 0.75$)	individual feedwater compressibility coefficient (ω = feedwater)
WRK-I	0.5	74	-25	-10
	1.0	47	-10	-2
	1.6	16	1	4
After	0.5	70	-13	-18
Ozonation	1.0	54	-27	-29
	1.6	25	-10	-11
After BACF	0.5	57	21	-13
	1.0	36	7	-9
	1.8	3	7	4
After SSF	0.5	64	1	9
	1.0	55	-24	27
Tap water	0.5	72	-19	10
	1.0	38	1	-5
	1.5	20	-3	1
diluted canal	0.6	63	-8	-4
water (10%)	1.0	30	18	-26
	1.5	16	4	-7

As expected from the tap water results, a large error was found when the MFI-UF value was not corrected for cake compression effects which was highest for the lowest applied pressure of 0.5 bar ranging from 57-74% for the feedwater examined. When the global compressibility coefficient was applied the maximum difference between the corrected and measured MFI-UF value at the reference pressure of 2 bar was 27% which was found for the WRK-I feedwater after ozonation measured at 1 bar. For 76% of the feedwater measurements, the global compressibility coefficient gave within ±20% of the true MFI-UF value experimentally determined at 2 bar. It was expected that employing the individual compressibility coefficient for a feedwater would give the most accurate means for pressure correction of the MFI-UF. This was indeed the case and 82% of the corrected MFI-UF feedwater measurements were within ±20% of the true MFI-UF value at 2 bar. However, this only represented a 6% improvement in pressure compression compared to the use of the global compressibility coefficient. In some cases e.g. the WRK-I water after ozonation and biological activated carbon filtration, the global compressibility coefficient gave a better approximation to the real MFI-UF value at reference conditions than the individual compressibility coefficient. This is expected to be coincidental as the individual compressibility coefficient must give a better fit. For these feedwater the compressibility coefficient should be redetermined.

6.4.5   Application of the MFI-UF Test as a Water Quality Monitor

As suggested in section 6.4.1.1. due to the long period of stability in the MFI-UF test, it could be applied as a monitor to detect changes in feedwater quality over time when operated with extended filtration. In Figure 6.10 the MFI-UF over time for two feedwater interchanged over time is presented. Initially, tap water was applied and the MFI-UF value stabilised after approximately 8 hours. Filtration was continued to 16.5 hours at which point the feedwater was changed on line to a 15% canal water dilution. Almost immediately the MFI-UF test responded to a change in feedwater quality and the MFI-UF value increased by 40% within 90 minutes. However, the MFI-UF value did not stabilise over time, instead it continuously increased over time. Therefore, an MFI-UF value could not be determined which could be used to characterise the fouling potential of the diluted canal water. At the time the feedwater was changed the MFI-UF was close to 12 000 s/l². In section 6.4.1.2. the MFI-UF value for a 15% canal water dilution was 11 000 s/l². The canal water used in this experiment was highly variable and this is expected to be only an indication of the MFI-UF value expected for this experiment.

It was expected that a distinct second cake layer with a structure and composition related to the nature of the particles present in the diluted canal water would be deposited directly on top of the cake formed by tap water filtration. The continuous increase of the MFI-UF value for this period may be due to the capture of smaller particles present in the diluted canal water in the tap water cake i.e. depth filtration is occurring. When particles with highly variable dimensions are present the cake may act as a filter itself and particles will penetrate and deposit within the cake [7]. This process will cause an increase in the specific resistance and the fouling index I will become time dependent [7].

At 65 hours the feedwater was changed back to tap water. Once again the same phenomena was observed whereby, the MFI-UF value registered a change. Although it was not as dramatic as when the feedwater was changed to the 15% canal water dilution. In this case the MFI-UF value only changed by 6% after 90 minutes. However, although the MFI-UF value began to decrease, the MFI-UF value did not stabilise. It was originally expected that a sandwich like structure of distinctive cake layers for each feedwater would be deposited. The fact that the MFI-UF value decreased indicated that an additional cake layer was indeed deposited. However, even after a further 40 hours of tap water filtration the MFI-UF value did not return to the original value of

3200 s/l² measured earlier for tap water. (The MFI-UF determined for tap water was found to be more consistent over time by the authors than canal water). This may be due to depth filtration throughout the previously deposited cake layers as mentioned above which may have been combined with cake compression which would increase the MF-UF value. Thus, the MFI-UF value obtained represents the overall result of these possible effects. Similar results to that presented in Figure 6.10 were found in other experiments when the feedwater was changed on line.

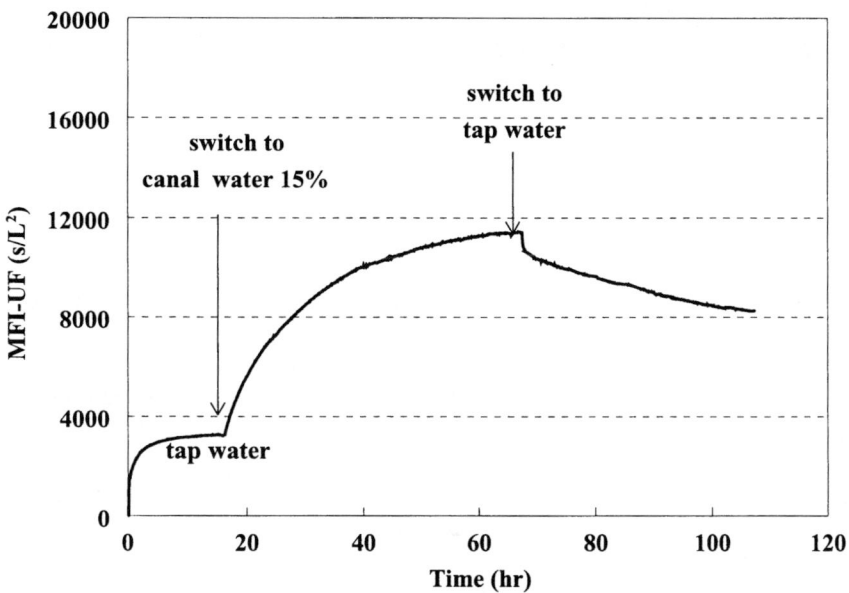

Figure 6.10: The MFI-UF determined over time where the feedwater was interchanged on line in the following sequence; tap water, 15% canal water dilution, tap water.

Another cause of the gradual increase or decrease observed in response to a feedwater change may be attributed to the method by which the MFI-UF is calculated and the accuracy of the flow meter. The MFI-UF is calculated from the gradient of the t/V versus V graph using cumulative time and volume data which contains an inherent "history effect". In long tests the impact of a change is adsorbed by the preceding data. Calculation of the MFI-UF using the first derivative (dt/dv vs. V) would remove this effect. However, as mentioned in Chapter 5 section 5.4.4. attempts to calculate the second derivative were unsuccessful. This was attributed to disturbances in the manual operation of pressure valves as the system stabilises at the start of filtration, combined with increasing fluctuations in the raw data due to the sensitivity of the flow meter as the flow decreased over time towards the end of filtration. Improvements to the MFI-UF equipment are expected to resolve the problem of data collection. The first (and second) derivative method to calculate the MFI-UF may then show instant changes in the MFI-UF in response to changes in feedwater quality over time and warrants further research.

In summary, at present the MFI-UF test cannot be applied to quantify the fouling potential of a feedwater with extended filtration i.e. operate as a monitor. However, it can be used as demonstrated in this chapter, as a test to characterise the fouling potential of a single given feedwater type and to register a change in feedwater quality e.g. failure of a pretreatment system preceding a membrane installation.

6.5   Conclusions

Cake filtration was shown to occur for high and low fouling feedwater by the stability of the MFI-UF value over time using the polyacrylonitrile 13 kDa membrane in the MFI-UF test. The MFI-UF value was found to be stable with prolonged filtration, up to circa 60 hours for WRK-I water. Cake filtration in the existing $MFI_{0.45}$ test occurred for a considerably shorter period of time circa 10 minutes for WRK-III water.

The MFI-UF index was demonstrated to be linear with particulate concentration for all the feedwater tested and at a range of applied test pressures which further supports that cake filtration is occurring in the MFI-UF test.

Manufacture of the PAN 13 kDa membrane within and between batches was found to be sufficiently reproducible to lead to a reproducible MFI-UF for the feedwater tested. The membrane was found to be reusable only with chemical cleaning to 100% clean water flux restoration in the MFI-UF test. The resultant MFI-UF value was then found to be reproducible with repeated testing.

The PAN 13 kDa membrane resistance was found to be temperature independent. In addition, the specific cake resistance measured in the fouling index I after correction for feedwater viscosity was found to be temperature independent for the 15-31 °C temperature range examined. Therefore, temperature correction in the MFI-UF test employing only a feedwater viscosity correction to the reference temperature of 20°C is sufficient.

The observed compaction of the PAN 13 kDa membrane with increasing applied pressure (0 to 2 bar range) was minor at 8% and does not have to be corrected for. Whereas, the specific cake resistance measured in the fouling index I (and hence the MFI-UF) was found to increase significantly with an increase in applied pressure due to cake compression effects. The compressibility coefficient determined for a range of feedwater was *circa* 0.51-0.85. The average of these individual feedwater compressibility coefficients (0.75) was used as a global compressibility coefficient to account for cake compression in the pressure correction factor. This resulted in 76% of the MFI-UF measurements within ±20% of the true MFI-UF value experimentally determined at the reference pressure of 2 bar. While the use of the individual feedwater compressibility coefficients, was marginally better with 82% of the MFI-UF measurements within ±20% of the true MFI-UF value.

At present the MFI-UF test can be used to characterise the fouling potential of a single given feedwater type and to register a change in feedwater quality. However, the MFI-UF test can not be applied to quantify the fouling potential of a variable feedwater over time i.e. operate as a monitor, as the resultant MFI-UF value may be due to the combination of depth filtration and compression effects with cake filtration. Moreover, the gradual increase or decrease in the MFI-UF value over time in response to a change in feedwater, may be due to the history effect in the calculation of the MFI-UF via the t/V vs V plot. More accurate measurement of time and volume is expected to resolve this problem and warrants further research.

Symbols

A	cross section area/membrane surface area [m^2]
A_o	reference surface area of a 0.45μm membrane filter [13.8×10^{-4} m^2]
C_b	concentration of particles in feedwater [kg/m^3]
CWF	clean water flux at 1 bar [l/m^2 h]
d_p	particle diameter [m]
h	membrane compaction coefficient [-]
I	index for the propensity of particles in water to form a layer with hydraulic resistance [1/m^2]
J	linear fluid velocity/permeate water flux [m^3/m^{2}s]
ΔP	applied transmembrane pressure [bar] or [N/m^2]
ΔP_c	pressure drop over the cake [bar] or [N/m^2]
ΔP_o	reference applied transmembrane pressure [2 bar]
Q	flow at temperature T [l/h]
R	resistance to filtration [1/m]
R_b	resistance due to blocking [1/m]
R_c	resistance of the cake [1/m]
R_m	membrane resistance [1/m]
R_{mo}	membrane filter resistance at zero compressive pressure [1/m]
r_p	diameter of particles forming the cake [m]
t	filtration time [s]
T	temperature [°C]
V	filtrate volume [m^3]

Greek Symbols

α	(average) specific cake resistance [m/kg]
α^1	local value of specific cake resistance [m/kg]
α_0	constant [m/kg]
ε	cake/membrane porosity [-]
η	fluid viscosity [Ns/m^2]
η_{20}	water viscosity at 20°C [Ns/m^2]
η_T	water viscosity at temperature (T) [Ns/m^2]
ρ_p	density of particles forming the cake [kg/m^3]
ω	compressibility coefficient[-]
τ	tortuosity of membrane pores []
Δx	membrane thickness [m]

Abbreviations

CWF   clean water flux at 1 bar [l/m^2 h]
$MFI_{0.45}$ Modified Fouling Index using a 0.45μm microfilter
MFI-UF Modified Fouling Index using an ultra filtration membrane
SDI   Silt Density Index

References

1.   J.C. Schippers and J.Verdouw, The Modified Fouling Index, a method of determining the fouling characteristics of water, Desal., 32, (1980), 137-148.
2.   W.K. Lewis and C. Almy, J. Ind. Eng.Chem. 4 (1912), 528.
3.   R.J. Wakeman, On the use of compressibility coefficients to model cake filtration processes, The Second World Filtration Congress, (1979) ,57-65.
4.   P.C. Carmen, Fundamental principles of industrial filtration, Trans. Inst. Chem. Engrs. 16, (1938), 168-188.
5.   B. F. Ruth, Studies in filtration, Ind. Eng. Chem., Vol. 27, 6 (1935), 708-723.
6.   P.M. Heertjes, Filtration, Trans.Inst.Chem.Engrs, 42 (1964), T266-274.
7.   J.C. Schippers, 1989 PhD Thesis ISBN 90-9003055-7, Vervuiling van hyperfiltratiemembranen en verstopping van infiltratieputten, Keuringsinstituut voor Waterleiding-artikelen Kiwa N.V., Rijswijk, The Netherlands.
8.   T. Kawakatsu, S. Nakao, and S. Kimura, Effects of size and compressibility of suspendedparticles and surface pore size of membrane on flux in cross flow filtration, J. Mem. Sci. (1993) 173-190.
9.   S. Chellam and J. G. Jacangelo, Existence of critical recovery and impacts of operational mode on potable water microfiltration, J. Env. Eng, 12 (1998).
10.   I. H. Huisman, B. Dutre, K. M. Persson and G. Tragardh, Water permeability in ultrafiltration and microfiltration, Viscous and Electroviscous effects, Desal., 113, (1997), 95-103.
11.   K. M. Persson, V. Gekas and G. Tragardh, Study of membrane compaction and its influence on ultrafiltration water permeability, J. of Mem. Sci, 100, (1995), 155-162.
12.   P.C. Carmen, Fluid Flow through Granular Beds, Trans. Inst. Chem. Engrs. 15, (1937), 150-166.
13.   F.M. Tiller, J. R. Crump and F. Ville, Filtration theory in its historical perspective, a revised approach with surprises, The Second World Filtration Congress, (1979),1-13.
14.   H. Banda and E. Forssberg, The effects of solid concentration, particle size and filtration pressure on filter cake resistance, Aufbereitungs-Technik, 9 (1987), 513-518.
15.   K. Rietma, Stabilizing effects in compressible filter cakes, Chem Eng. Sci, 2 (1953), 88-94.

7

Applications of the MFI-UF

Chapter 7 is based on: Modified Fouling Index$_{\text{-Ultrafiltration}}$ to Compare Pretreatment Processes of Reverse Osmosis Feedwater" by S.F.E. Boerlage, M.D. Kennedy, M. p. Aniye, E. M. Abogrean D. E.Y. El-Hodali, Z. S.Tarawneh and J.C. Schippers. Published in *Desalination*, Vol. 131 pp. 201-214 (2001).

Contents

Abstract

The MFI-UF was developed to include smaller colloidal particles not measured in the existing SDI and $MFI_{0.45}$ fouling indices. This research investigates its application to measure and predict the particulate fouling potential of reverse osmosis (RO) feedwater. MFI-UF measurements were carried out under constant pressure filtration at the IJssel Lake and River Rhine RO pilot plants of the influent feedwater and after pretreatment processes e.g. coagulation, sedimentation, conventional filtration, ultrafiltration etc. Using the MFI-UF results, the pretreatment efficiency was evaluated and a comparison made with the SDI, $MFI_{0.45}$, and particle counts. The MFI-UF of the influent feedwater was circa 700-2400 higher than the corresponding $MFI_{0.45}$ and SDI, due to the retention of smaller particles. A pretreatment efficiency of $\geq 80\%$, was found by MFI-UF at both plants. For the larger particles the $MFI_{0.45}$ gave a 90-$\approx 100\%$ reduction. Minimum predicted run times for a 15% flux decline from MFI-UF measurements were shorter than that observed at the IJssel Lake plant. This was most likely due to almost negligible particle deposition in the RO systems and/or particle removal from the cake formed under cross flow. Moreover, it was shown that cake resistance increased with ionic strength in MFI-UF tap water experiments and therefore, a correction of the MFI-UF index is required for salinity effects in RO concentrate. Finally, it was suggested that the MFI-UF be carried out under constant flux (CF) filtration to more closely simulate fouling in RO systems. Preliminary experiments were promising, the MFI-UF could be determined under CF filtration within ≈ 2 hours for the low and high fouling feedwater examined.

7.1   Introduction

There is a pressing need for a method which enables the measurement and prediction of the particulate fouling potential of feedwater in membrane filtration systems. This method can then be used to assess the effectiveness of a pretreatment step and monitor its performance over time. Water quality parameters such as turbidity, suspended matter and particle counting can be used to indicate the concentration of particles in a feedwater. In particular, particle counting has proven particularly successful in monitoring particle removal and the integrity of ultrafiltration membranes [1-5]. However, the aforementioned parameters cannot be used to estimate the particulate fouling potential of a feedwater arising from the deposition of particles on the membrane or spacer. Existing methods designed to measure the particulate fouling potential of RO and NF feedwater are; the Silt Density Index (SDI), and the Modified Fouling Index ($MFI_{0.45}$). In both tests the water is filtered through a $0.45\mu m$ microfiltration membrane in dead-end flow and at constant pressure. The SDI, however, is not based on a distinction between filtration mechanisms occurring during the test and is not proportional with particle concentration [6]. Therefore, it can not be used for predicting the rate of flux decline due to particulate fouling, making use of a mathematical model.

Conversely, the $MFI_{0.45}$ is based on cake filtration and can be used to model flux decline or pressure increase to maintain constant capacity in membrane systems assuming that particulate fouling of RO and NF membranes is dominated by cake filtration [7-9]. Moreover a linear relationship with particle concentration was proven [6]. Calculations based on the cake filtration model show that the measured $MFI_{0.45}$ values are far too low to explain the observed flux decline rates in practice [8]. Therefore, a $MFI_{0.05}$ using membranes with pores of $0.05\mu m$ was developed. However, it is likely that particles much smaller than $0.05\mu m$ which are not retained by the test membranes, are responsible for the observed flux decline rates [8,9]. Consequently, until now, there was no method available to measure and assess the fouling potential of these particles in a feedwater. Furthermore, the efficiency of various pretreatment processes *e.g.* coagulation, sedimentation, conventional filtration (slow sand filtration/rapid sand filtration), ultrafiltration, etc. in removing smaller colloidal particles and in reducing the particulate fouling potential of a feedwater could not be evaluated.

Therefore, the MFI-UF was developed (Chapter 5) as a new parameter to include the fouling potential of these smaller particles. The average smallest pore dimension of the ultrafiltration membrane was estimated by scanning electron microscopy to be 9 nm and thus larger and to some extent smaller particles are retained [10]. A linear relationship with particle concentration was demonstrated for the new MFI-UF index and cake filtration was proven to occur [10-12]. Consequently, the MFI-UF can also be applied in the MFI model developed by Schippers [7-9] for fouling prediction. This model enables the prediction of the time required for either (i) a per centage decline in the normalised flux or (ii) an increase in the applied pressure to occur in a membrane filtration system operating in constant pressure or flux mode, respectively.

In order to achieve more accurate prediction in these models, differences in the applied pressure, flux, and flow mode between the MFI-UF test conditions and the target membrane system e.g. reverse osmosis needs to examined. The existing $MFI_{0.45}$ and MFI-UF are conducted under a constant applied pressure of 2 bar while the low pressure brackish water membranes operate at 10-20 bar and sea water up to 90 bar. In the preceding chapter the MFI-UF was found to be pressure dependent, as was expected from previous research on the $MFI_{0.45}$ [9]. Whereby, compression of the cake occurred at a higher applied pressure, increasing the cake resistance and hence causing a faster flux decline. A method suggested by Schippers [9] for correction of the $MFI_{0.45}$ for pressure compression effects was verified for the MFI-UF within the operating pressures of the test (refer Chapter 6 section 6.4.4.2.). This method could be extended to correct the MFI-UF to the operating pressures of the target membrane system.

In addition, the flux applied in membrane systems e.g. reverse osmosis (RO) may be 10-1000 times lower than that in the MFI-UF test. From literature flux may also be expected to cause cake compression, and indeed Schippers found the $MFI_{0.45}$ was also dependent on flux [9]. Thus, cakes formed on RO membranes may be compressed to a lesser extent than those formed on the MFI-UF test membrane. Therefore, most likely the MFI-UF may require flux correction to that of the target system. Furthermore, most membrane installations work in the constant flux mode with increasing pressure to maintain flux. The MFI-UF determined in constant pressure mode may then be limited in simulating the cake resistance formed in constant flux mode due to the variation in flux over time during the determination in constant pressure mode. Thus, it may be more accurate to determine the MFI-UF in constant flux mode with the transmembrane pressure becoming a dependent variable. In this way the particulate flux to the membrane remains constant.

Another consideration in modelling particulate fouling is the flow mode under which the membrane system operates, either dead end or cross flow mode. Most UF systems operate in dead end mode while most RO systems operate in cross flow mode. In the former case all the water passes the membrane while in the latter mode, the feedwater flows tangentially across the membrane surface with only a part of the feedwater passing through the membrane. Ideally, in the cross flow mode the convection of particles to the membranes is equally balanced by back transport of the particles away from the membrane. Thus, no cake forms and the permeate flux can be maintained. In practice however, a thin cake is often formed, whereby, some of the particles deposit on the membrane despite the cross flow velocity [13-15]. The $MFI_{0.45/0.05}$ and MFI-UF operate in dead end mode. To account for differences between dead end and cross flow filtration the MFI model incorporates a particle deposition factor which can be experimentally determined.

The research presented in this chapter is focussed on the application of the MFI-UF test as a tool to measure and predict the particulate fouling potential of RO/NF feedwater. The MFI-UF of conventionally pretreated RO feedwater at two RO pilot plants, after extended pretreatment processes in surface water treatment such as ultrafiltration, will be measured in the existing constant pressure mode. The efficiency of pretreatment processes on particle removal will be assessed applying the MFI-UF and a comparison made with the existing SDI and $MFI_{0.45}$ indices and particle counts where possible. Furthermore, particle counting will be used to determine the proportion of particles retained by the MFI-UF test membrane. The MFI-UF measured for the RO feedwater after pretreatment will be applied in the fouling prediction model, based on cake filtration, developed by Schippers [7-9] and compared to that observed in practice, assuming all particles are deposited in the RO system. The particle deposition factor will then be determined at the RO pilot plants. Finally, preliminary experiments to determine the MFI-UF of a low and high fouling feedwater in the constant flux (CF) mode, the effect of applied flux on the index and the linearity of the index with colloidal concentration in the CF mode will be investigated.

7.2   Background

7.2.1   The Modified Fouling Index

7.2.1.1   The MFI Determined in Constant Pressure Mode

The Modified Fouling Index (MFI) is based on cake filtration, whereby, particles are retained on the membrane during filtration by a mechanism of surface deposition. The MFI can be measured at constant pressure using the well known cake filtration equation where the MFI is defined as the gradient of the linear region found in the plot of t/V vs V (refer Chapter 5 section 5.2.1):

$$\frac{t}{V} = \frac{\eta\,R_m}{\Delta P\,A} + \frac{\eta\,I}{2\,\Delta P\,A^2}\,V$$

$$\uparrow$$
$$MFI$$

...(7.1)

where V is filtrate volume, t filtration time, ΔP the applied transmembrane pressure, η the water viscosity and A the membrane surface area. The MFI is calculated at standard reference values of ΔP (2 bar), η ($\eta_{20°C}$) and A the surface area of the 0.45 μm test membrane (13.8×10^{-4} m^2). I in Equation 7.1 refers to the fouling index which is a measure of the fouling potential of the feedwater. The fouling index I is defined as the product of α the specific resistance of the cake deposited and C_b the concentration of particles in the feedwater (Equation 7.2). An advantage of using the fouling index I is that in most cases it is impossible to determine C_b and α accurately.

$$I = \alpha\,C_b$$

...(7.2)

Equation 7.2 is for an incompressible cake. However, as discussed in Chapter 6 (section 6.2.2) the MFI is also valid for compressible cakes as when the pressure drop over the cake ΔP_c is dominant, a linear relationship between t/V versus V will appear and the MFI value can then be determined. The fouling index I taking into account cake compression is defined as follows:

$$I = \alpha_o\,\Delta P_c^{\omega}\,C_b$$

...(7.3)

where ω is the compressibility coefficient and α_o a constant. For incompressible cakes, ω is zero and the higher the compressibility coefficient the more compressible the cake. Substitution of the Carmen-Kozeny relationship (refer Equation 5.5) for the specific resistance of a cake composed of spherical particles in the MFI gives:

$$MFI = \frac{\eta_{20^\circ C} \, 90(1 - \varepsilon)C_b}{\rho_p \, d_p^2 \varepsilon^3 \, \Delta P_o \, A_o^2} \qquad \qquad ...(7.4)$$

Thus, the MFI is a function of the dimension and nature of the particles forming a cake on the membrane, and directly correlated to the concentration of particles in a feedwater. At the fixed reference values of ΔP_o, $\eta_{20^\circ C}$ and A_o, the MFI can be used to characterise and compare the fouling potential of feedwater containing particles [6-9].

7.2.1.2   The MFI Determined in Constant Flux Mode

The MFI can also be determined under constant flux filtration. In this mode of filtration the permeate flux is kept constant and the pressure is increasing to maintain a constant flux. Taking the standard equation (refer Chapter 5 Equation 5.1) describing the flux through a membrane:

$$\frac{dV}{A \, dt} = \frac{\Delta P}{\eta \, (R_m + R_c)} \qquad \qquad ...(7.5)$$

and substituting J for flux and R_c by:

$$R_c = \frac{V}{A} \times I \qquad \qquad ...(7.6)$$

the following equation is obtained:

$$J = \frac{\Delta P}{\eta \, (\frac{V}{A} I + R_m)} \qquad \qquad ...(7.7)$$

rewriting V/A as Jt and rearranging gives:

$$\Delta P = J \eta \, R_m + J^2 \eta \, I \, t \qquad \qquad ...(7.8)$$

The fouling index I can then be determined from the slope of the linear region in a plot of ΔP versus time which corresponds to cake filtration or by manipulation of Equation 7.8. The MFI can be calculated using I (from Equation 7.8) for standard reference conditions as follows:

$$MFI = \frac{\eta_{20^\circ C} I}{2 \, \Delta P_o \, A_o^2} \qquad \qquad ...(7.9)$$

7.2.2   MFI Fouling Prediction Models

The MFI models to predict fouling developed by Schippers [7-9] are based on the assumption that particulate fouling on the surface of reverse osmosis (or nanofiltration) membranes can be described by the cake filtration mechanism. The relationship between the MFI measured for a feedwater and the flux decline predicted for a reverse osmosis system are derived below. When scaling, adsorptive blocking and biofouling do not contribute to the fouling observed, the flux through a RO membrane is:

$$J_r = \frac{dV_r}{A_r \, dt_r} = \frac{\Delta P_r}{\eta_r (R_{mr} + R_{cr})} \qquad ...(7.10)$$

where the subscript r indicates that the parameter refers to filtration through a RO membrane. The resistance of the cake formed during constant pressure filtration provided the retention of particles is constant is:

$$R_{cr} = \frac{V_r}{A_r} \times I_r \qquad ...(7.11)$$

Substituting J_o for time $t = 0$ and J for time $t = t$, the relative flux decline may be defined as:

$$\Delta J_r = \frac{J_{or} - J_r}{J_{or}} = 1 - \frac{J_r}{J_{or}} \qquad ...(7.12)$$

The relationship between I_r and I (from the MFI measurement) is assumed as:

$$I_r = \Psi \Omega I \qquad ...(7.13)$$

where the cake ratio factor (ψ) accounts for differences between the cake deposited on the MFI membrane and that deposited on the RO membrane and the particle deposition factor (Ω) represents the ratio of the particles deposited on the RO membrane to that present in the feedwater. The particle deposition factor is calculated from the relation between the MFI of the concentrate at recovery Y of the RO system and the MFI of the feedwater as follows:

$$\Omega = \frac{1}{Y} + \frac{MFI_{concentrate}}{MFI_{feed}} \left(1 - \frac{1}{Y}\right) \qquad ...(7.14)$$

Which is calculated from the mass balance of:

$$Q_f MFI_{feed} = \Omega Q_p MFI_{feed} + Q_c MFI_{concentrate} \qquad ...(7.15)$$

where Q is the flow of feed (f), concentrate (c) and permeate (p). Combining Equation 7.5. with Equations 7.9 to 7.13 gives:

$$t_r = \frac{\eta_{20°C} \, \Delta P_r}{\eta_r \, J_o^2 \, \Psi \Omega \, (MFI) \, 2\Delta P_o \, A_o^2} \times \frac{\Delta J (2 - \Delta J)}{2(1 - \Delta J)^2} \qquad ...(7.16)$$

where t_r is the time in which the flux of a RO membrane has decreased by a factor (ΔJ).

Similarly, for a RO system operating under constant flux filtration, the time required for a increase in pressure ΔP_r to occur can be predicted by manipulation of Equation 7.7 applied to a RO membrane:

$$t_r = \frac{\Delta P_r - \Delta P_{or}}{J^2 \eta I_r} \qquad \qquad ...(7.17)$$

Assuming I_r is related to I from the MFI measurement (Equation 7.13) and substituting in Equation 7.9 with rearrangement, I_r can be defined as:

$$I_r = \frac{2 \, MFI \; \Psi \, \Omega \, \Delta P_o \, A_o^2}{\eta_{20°C}} \qquad \qquad ...(7.18)$$

Combining Equation 7.17 and 7.18 gives the time t_r period in which the pressure of a RO system has increased from an initial operating pressure of ΔP_{or} to ΔP_r.

$$t_r = \frac{\eta_{20°C}(\Delta P_r - \Delta P_{or})}{\eta_r \, J_o^2 \; \Psi\Omega(MFI)\, 2\Delta P_o \, A_o^2} \qquad \qquad ...(7.19)$$

7.3   Materials and Methods

7.3.1   Feedwater

Feedwater tested included; Delft tap water, Delft canal water and RO feedwater after various extended pretreatment processes and membrane concentrate at the RO pilot plants of Amsterdam Water Supply (AWS) Company and Provincial Water Supply Company of North Holland (PWN). Details of the pretreatment schemes and the RO pilot plants are given below:

The AWS RO pilot plant uses conventionally pretreated (coagulation, sedimentation, and filtration) River Rhine water (WRK-I) which is further treated in the RO I pretreatment system by ozonation, biological activated carbon filtration and slow sand filtration. Acid is added to the RO feedwater and at the time of measurement antiscalant (Perma Treat 191) was also added. However, it was not possible to measure the feedwater after either acid or antiscalant addition. The RO pilot plant at AWS comprises a three stage system and at the time the membrane concentrate was tested the RO system operated at 85% recovery.

The PWN RO pilot plant applies IJssel Lake water which has undergone conventional pretreatment; storage reservoir, coagulation, sedimentation, rapid sand filtration (WRK-III) and is then ultrafiltered through a membrane with a *maximum* MWCO rating of 150-200 000 Da (X-flow hollow fibre polyethersulphone) followed by hydrochloric acid and antiscalant (Perma Treat 191) addition prior to reverse osmosis. At the time of testing the PWN RO pilot plant operated three *two stage* RO pilot plants with antiscalant addition (1) HT (2) HA and (3) HC at a recovery of 80% and a *single stage* RO pilot plant with no antiscalant addition (ii) HD at a recovery of circa 50% recovery.

Details of the membranes and the date of installation for the AWS and PWN RO pilot plants are summarised in Table 7.1, dates of MFI-UF measurements and cleaning carried out of the RO systems are included.

Table 7.1:   Membranes and installation dates at the AWS and PWN RO pilot plants and the date of membrane cleaning and of MFI-UF measurements.

RO pilot Plant	Membrane Brand Date installed	Dates of Cleaning	Dates of MFI-UF measurement
AWS RO I	Fluid Systems 4821 ULP March 1997	10th June 1998 28th September 1998 8th October 1998 20th January 1999	1-14th September 1998 8-23rd February 1999 29th March 1999 6th April 1999
PWN HT (2 stage)	Dow Filmtec April 1998	not cleaned since January 1998	
HA (2 stage)	Hydranautics ESPA July 1997	never been cleaned (at the time of sampling)	1-November- 4th December 1998
HC (2 stage)	Hydranautics ESPA March 1998	never been cleaned (at the time of sampling)	
HD (single stage) no antiscalant addition	Hydranautics ESPA March 1998	preventative cleaning every 14 days	20th November 1998 27th November 1998

7.3.2   MFI-UF Determination

All feedwater were applied on line to the MFI-UF equipment (refer Figure 5.1 in Chapter 5) with the exception of the diluted canal water which was applied in the same way as described in Chapter 6 section 6.3.5. The MFI-UF was determined either in the constant pressure or constant flux mode as described below. The detection limit of the MFI-UF test is circa 80-100s/l^2 and the accuracy is expected to be ±10%. Three MFI-UF apparatuses were used in this study.

After a MFI-UF test the membranes were backwashed (1.5 bar) with RO permeate (ambient temperature) for 15 minutes followed by a chemical cleaning using a 200ppm sodium hypochlorite solution which was recirculated for 1 hour at 1 bar. Sodium hypochlorite was prepared from analytical grade reagent and RO permeate. To remove any chemicals the membranes were backwashed again before the clean water flux was measured. Membranes were cleaned until ≈100% clean water flux restoration.

7.3.2.1   Constant Pressure Mode

The MFI-UF of a feedwater was determined in constant pressure mode as described in Chapter 5.3.2. The software program calculated the MFI-UF in constant pressure mode according to Equation 7.20 normalised to the standard reference conditions of the existing $MFI_{0.45}$ namely; temperature of 20°C, transmembrane pressure of 2 bar (ΔP_o) and surface area of the MFI $_{0.45}$ microfiltration membrane (A_o):

$$MFI - UF = \frac{\eta_{20°C}}{\eta_T} \left(\frac{\Delta P}{\Delta P_o} \right)^{1-\omega} \left(\frac{A}{A_o} \right)^2 \frac{d\frac{t}{V}}{dV} \qquad ...(7.20)$$

and is therefore directly comparable with the $MFI_{0.45}$. The global pressure correction (ω) value of 0.75 determined in Chapter 6 was applied to correct for cake compression effects. The MFI-UF reported represents the average calculated over the stable region (corresponds to cake filtration) of the MFI-UF over time in a single test (refer Chapter 5) .

7.3.2.2   Constant Flux Mode

To determine the MFI-UF of a feedwater in the constant flux mode, the pressure reducing valve of the MFI-UF equipment (refer Figure 5.1 in Chapter 5) was manually adjusted to maintain the required applied flux at a constant value. The fouling index I was calculated at time (t) by manipulation of Equation 7.8 which gives:

$$I = \frac{\Delta P - J\eta_{20°C}R_m}{J^2\eta_{20°C}t} \qquad ...(7.21)$$

Where J is the applied flux corrected to 20°C and R_m is the membrane resistance at the beginning of the test. The fouling index I was input into the Equation 7.9 to determine the MFI-UF at standard reference conditions.

7.3.3   Particle Counting

A particle counter (Particle Measuring Systems HSLIS model M-50) was used to count particles in the IJssel Lake feedwater after (i) UF (ii) acid and antiscalant addition and (iii) filtration through the MFI-UF test membrane. The counter operates on-line and counts particles per mL in four channel sizes (μm); 0.05-0.10, 0.10-0.15, 0.15-0.20, and ≥0.20. The aforementioned feedwater were fed to the particle counter for one hour prior to measurement and the average taken of a minimum of seven measurements. Particle counts measured for the RO permeate was used to determine the background noise of the instrument and subtracted from the measurements reported.

7.3.4   Flux Decline Prediction at the RO Pilot Plants

The AWS and PWN pilot plants operate at a constant flux, with the applied pressure varying between 8 bar in summer to 10 bar in winter. To simplify prediction of particulate fouling in the RO systems, it was assumed that both the RO systems operated at a constant pressure of 10 bar (ΔP_r) for a feedwater temperature of 20°C (i.e. η_r is for 20°C) with an initial flux of 25 l/m²/hr (J_o). The time t_r for a 15% decline in flux (ΔJ) to occur was calculated by Equation 7.16 by inputting the MFI-UF measured of the RO feedwater. The MFI-UF value was corrected to a pressure of 2 or 10 bar applying either the global or individual pressure compressibility coefficient in Equation 7.20.

7.3.5   Determination of Particle Deposition Factor

The MFI-UF of the feedwater and the concentrate were determined in the constant pressure mode at the AWS and PWN RO pilot plants. The MFI-UF was corrected to the desired pressure drop of the RO system using either the individual or global compressibility coefficient ($\omega = 0.75$) in Equation 7.20. The particle deposition factor was calculated using Equation 7.14. Experiments to determine the particle deposition factor were carried out at the PWN RO pilot plants in November 1998 and at the AWS pilot plant from 29th March to 6th April 1999.

7.3.6   Determining the Ionic Strength Dependency of the MFI-UF

Sodium chloride predissolved in circa 10 litres of tap water was added to a 2m³ tank of tap water to prepare solutions of 0.032, 0.064, 0.1, 0.15 and 0.2 M ionic strength. The contents were allowed to equilibrate to ambient temperature prior to use and were stirred during the experiment. The reference pressure of 2 bar was not used due to limitations in pump capacity at the time of experiments. Therefore, the MFI-UF was determined at an applied transmembrane pressure of 1.5 bar and corrected to the reference pressure of 2 bar as per Equation 7.20 where ω is 0.75.

7.3.7   Linearity Experiments

Dilutions of tap and canal water were prepared as described in Chapter 6 section 6.3.5. Canal water was pretreated prior to dilution, by filtering through a 60μm nylon mesh to remove large contaminants. Experiments for all feedwater were carried out under a constant applied flux of circa. 75 l/m²/hr as described in section 7.3.2.2. One PAN 13 kDa membrane was used for all feedwater with chemical cleaning in between measurements to 100% CWF. Tap water experiments were carried out from 29th June to 18th July and the canal water experiments circa one month later from 26th July to 6th August 1999.

7.4   Results

7.4.1   MFI-UF to Assess Pretreatment Efficiency

7.4.1.1.   River Rhine Pilot Plant

The MFI-UF of the WRK-I influent feedwater and after various pretreatment steps, measured on two occasions at the River Rhine RO pilot plant, is presented in Figure 7.1. The WRK-I feedwater still contains particles after conventional pretreatment as the MFI-UF measured in September 1998 was 3600 s/L^2 and in February 1999, 2800s/L^2. The MFI-UF value and hence the particulate fouling potential decreased with increasing levels of pretreatment (refer Figure 7.1) due to the removal of particulates from the feedwater.

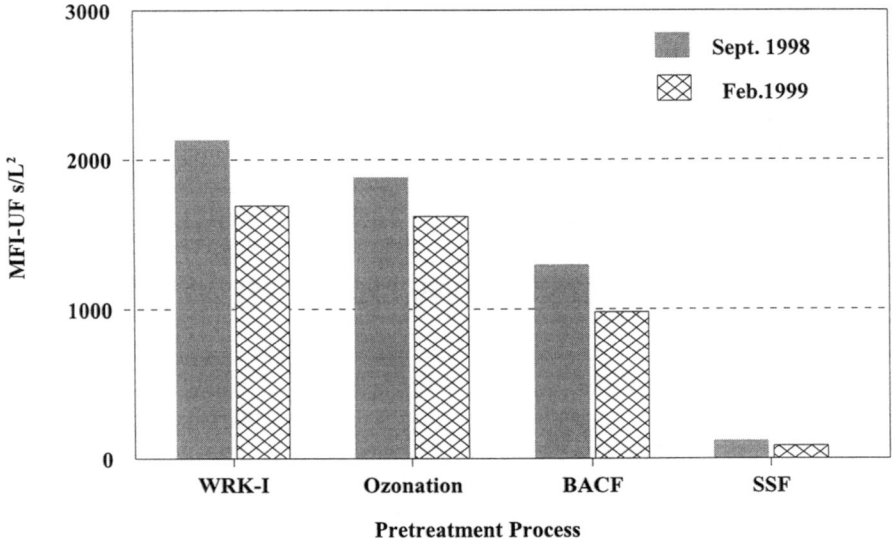

Figure 7.1:   Effect of pretreatment processes on the MFI-UF at the River Rhine RO pilot plant which applies conventionally pretreated water (WRK-I) followed by ozonation, biological activated carbon filtration (BACF) and slow sand filtration (SSF).

$MFI_{0.45}$ and SDI measurements available from the AWS pilot plant for the corresponding time (or closest available) period are reported in Table 7.2. On both occasions the $MFI_{0.45}$ of the WRK-I water was greater than 1 and according to pilot plant experience, $MFI_{0.45}$ values >1 *in the RO feedwater* resulted in RO plant run times of less than two weeks [16]. In comparison to the MFI-UF measured for the WRK-I water, both the $MFI_{0.45}$ and SDI are significantly lower, e.g. for the August measurement circa 700-2400 times lower, respectively. This is because smaller particles are not retained by the microfiltration test membranes and therefore, their fouling potential is not measured. Similarly to the MFI-UF, the $MFI_{0.45}$ of the WRK-I feedwater was lower in February compared to August. The opposite effect was found in the SDI measurements, whereby, the February measurement was circa. 5% higher than that of August. However, as in the case of the MFI-UF both the $MFI_{0.45}$ and the SDI decreased with pretreatment indicating a reduction in the fouling potential for particles of the size measured in these tests.

Table 7.2: The $MFI_{0.45}$ and SDI determined at the River Rhine RO pilot plant for the influent WRK-I water and after increasing levels of pretreatment; biological activated carbon filtration and slow sand filtration.

Feedwater	$MFI_{0.45}$ (s/L²)		SDI (-)	
	August [a]	February	August [a]	February
WRK-I	4.9	3.4	1.5	1.6
after BACF	2.1	1.8	1.4	0.9
after SSF	≤0.1 [b]	0.5	≤0.1 [b]	0.5

(a) - last measurement in August
(b) - below the detection limit

The efficiency of the various pretreatment processes in reducing the particulate fouling potential of WRK-I water as assessed by the MFI-UF, $MFI_{0.45}$ and SDI is given in Table 7.3. In general, the removal efficiency of the aforementioned tests is for particles larger or equal to the pore size of the membrane, e.g for the MFI-UF particles $\geq 13\,000$ dalton or as determined in Chapter 5 an estimated pore size of ≥ 9nm (corresponds to the limiting pore dimension measured). However, as discussed in Chapter 5, during blocking filtration the membrane pores may become partially blocked reducing their effective pore size and the retention of smaller particles is expected to increase. Moreover, the cake itself may retain particles acting as a second membrane with smaller pore interstices in the cake than the membrane. Furthermore, smaller particles may aggregate into larger particles by bridging. As a result of these mechanisms, particles up to ten times smaller than the membrane pore size rating may be retained in the test. Therefore, the particle size used henceforth of "$\geq 13\,000$ Da" for the MFI-UF and "$\geq 0.45\mu$m" for the $MFI_{0.45}$ and SDI, should be understood to include the smaller particles retained by these processes. Notwithstanding, the removal efficiency will be higher for the $MFI_{0.45}$ and SDI than the MFI-UF as it is expected that larger particles are more easily removed in the treatment process. While the removal efficiency of the MFI-UF includes the removal of these larger particles plus smaller particles which may or may not be removed from the feedwater.

The effect of ozonation, the first step in the pretreatment process, on the particulate fouling potential is difficult to predict. Ozonation will oxidise organic matter e.g. humic acids into lower molecular weight, more polar and more biodegradable substances [17]. Therefore, both the MFI-UF (and $MFI_{0.45}$) would be expected to increase due to the increased number of particles with a smaller particle diameter via Equation 7.4. Conversely, some particles may become so small that they can pass the interstices of the cake and the pores of the membrane, resulting in a decrease in the MFI-UF. Moreover, the natural organic matter "mantle" surrounding colloidal particles which stabilises the particle may be hydrophilic, and this mantle may be destroyed or reduced by ozonation [18]. This effect will lead to particle destabilization and aggregation and has been shown to occur particularly in the presence of metal ions such as calcium [18]. Thus, the particle size may increase and the MFI-UF value will decrease. It was expected that the net result of the aforementioned effects may cancel giving little or no change in the MFI-UF after ozonation of the WRK-I water. This was indeed the case for February 1999 where the MFI-UF decreased by only ≈5% after ozonation (refer Table 7.3). A larger reduction (10%) in the MFI-UF was found earlier in September 1998. However, given the accuracy of the MFI-UF measurement, circa ≈10%, this difference is considered marginal. The efficiency of the ozonation step cannot be examined with the $MFI_{0.45}$ and the SDI as the feedwater is not routinely measured after ozonation.

Table 7.3:   The removal efficiency per pretreatment step and overall treatment removal efficiency (%) at the River Rhine RO Pilot Plant as assessed by the MFI-UF, $MFI_{0.45}$ and SDI tests.

Feedwater	Removal Efficiency of Pretreatment Step (%)					
	MFI-UF (s/L^2)		$MFI_{0.45}$ (s/L^2)		SDI (-)	
	September	February	August	February	August	February
after ozonation	10	5	-	-	-	-
after BACF	30	40	60	50	10	40
after SSF	90	90	≈100	70	>90	40
Overall treatment removal efficiency (WRK-I to SSF)	90	90	≈100	90	>90	70

- not routinely measured at the pilot plant

A more significant reduction in the particulate fouling potential measured by the MFI-UF was found in the ozonated feedwater after biological activated carbon (BACF) filtration, 30 and 40% for September and February, respectively. The reduction in the MFI-UF is due in part to the indirect effect of ozonation, whereby, the smaller more biodegradable organic molecules may be removed by bacteria growing on the activated carbon and by the direct chemisorption of organic matter onto the activated carbon. The BACF filter was not backwashed during the time of measurements nor the activated carbon regenerated. Therefore, the efficiency of this step is expected to be constant and the 10% difference in efficiency is considered marginal and within experimental error. The total removal of particulates in the BACF filter is not expected as a portion of the inorganic particles will pass through the filter due to the rather large size of the activated carbon particles (oblong in shape: 3.1×0.8mm).

The removal efficiency of the combined ozonation and BACF steps using the $MFI_{0.45}$ test was 60 and 50% for August and February, respectively. This indicates that circa 55% of the particles removed, principally in the BACF step by chemisorption and bacterial metabolism of organic matter as discussed above, are "≥0.45µm". In addition, some inorganic particles in this size range might be removed by mechanical sieving. In contrast, the SDI gave an unexpectedly low removal efficiency of 10% for the August measurement, as the SDI removal should be higher than the corresponding MFI-UF. While for the February measurement, the SDI gave a 4 times higher removal efficiency for the same treatment step. As the SDI does not make any distinction between filtration mechanisms, one measurement may correspond predominantly to cake filtration and the other blocking filtration. Furthermore, there is no linear correlation between the SDI and the concentration of colloidal/suspended particles. Therefore, the SDI removal efficiencies have only limited value and consequently, the removal efficiencies observed for the $MFI_{0.45}$ and SDI do not correspond. The $MFI_{0.45}$ results are considered to be more reliable because of the theoretical basis of this test.

The final pretreatment step, slow sand filtration (SSF) was assessed by all three tests as the most successful in removing particulates from the feedwater. This is due to depth filtration through the sand grains (size 0.1-0.3mm) and their entrapment and metabolism in the biofilm layer ("Schmutzdeke"), which typically develops at the water sand interface of slow sand filters and which greatly aids the filtration process. Both MFI-UF measurements gave a 90% reduction in the fouling potential of the feedwater after SSF. Moreover, the MFI-UF measurements demonstrated that SSF consistently removed particles to an MFI-UF of $\leq 200s/l^2$, balancing out differences of the other pretreatment processes. In fact, the MFI-UF was very close to the detection limit of the test and thus the fouling potential of the RO feedwater was very low. In terms of the removal of particles "$\geq 0.45\mu m$", the $MFI_{0.45}$ test gave $\approx 100\%$ removal for the SSF step in August. The removal efficiency was lower (70%) than expected in February, which may be due to the accuracy of the measurement close to the detection limit ($0.1s/l^2$), as it should be higher than the MFI-UF. Nevertheless, the $MFI_{0.45}$ measured for the feedwater after SSF was very low, $0.1-0.5s/l^2$, on both occasions.

Both the MFI-UF and $MFI_{0.45}$ tests showed the pretreatment efficiency per step was similar when measured on the two occasions. As a consequence the overall removal efficiency achieved by the River Rhine pretreatment scheme for the RO feedwater was similar. The MFI-UF and $MFI_{0.45}$ results showed that $90-\approx 100\%$ of the particles "$\geq 13\,000$ kDa" and "$\geq 0.45\mu m$", respectively, are removed from the influent WRK-I feedwater. In comparison, the SDI gave variable results for the removal efficiency of a pretreatment step and for the overall treatment removal efficiency. This may be attributed to the limitations of the SDI test as mentioned above. The SDI results for the overall removal efficiency of the pretreatment scheme were similar in August (>90%) to the $MFI_{0.45}$ result ($\approx 100\%$), however, this is expected to be by coincidence as in February it was 20% lower than that found by the $MFI_{0.45}$ test.

7.4.1.2   IJssel Lake Pilot Plant

Results of MFI-UF measurements at the IJssel Lake pilot plant of the WRK-III influent feedwater and after various pretreatment steps are presented in Figure 7.2. The two MFI-UF measurements of the WRK-III feedwater were only two days apart (December 2nd and 4th) and the MFI-UF value was similar at $\approx 8500s/L^2$. This is circa 2.4-3.0 times higher than WRK-I (refer Figure 7.1). The lower quality of the WRK-III water after conventional treatment is due to the storage characteristics in the IJssel Lake and storage reservoir. The IJssel Lake is eutrophic, with a high concentration of nutrients, algae and bacteria etc. which release extracellular material which is particulate in nature. In addition, there is ground water seepage from neighbouring polders composed of peat layers. The incomplete mixing of the Lake leads to areas with a high localised concentration of organic (particulate) material and nutrients and hence the difficulty in treating the IJssel Lake water.

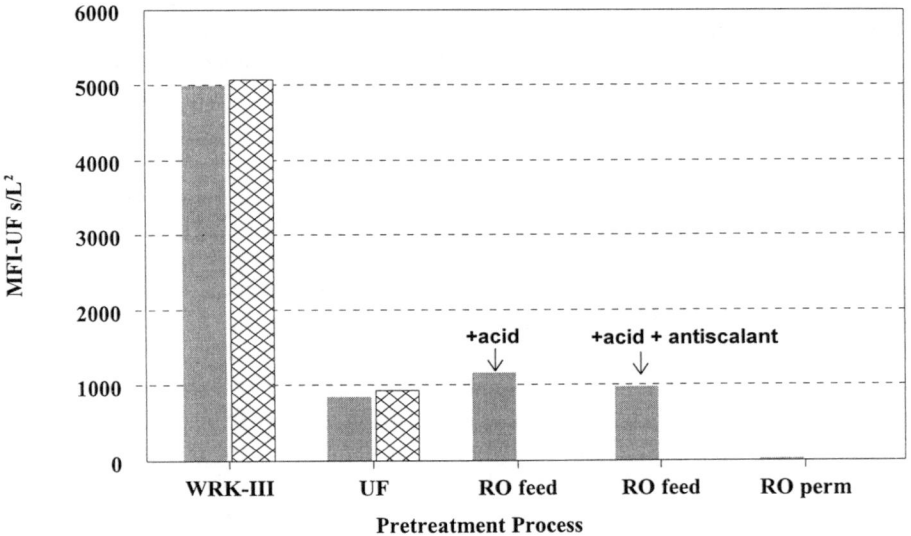

Figure 7.2:   The MFI-UF measured at the IJssel Lake RO pilot plant which applies conventionally pretreated water (WRK-III) followed by ultrafiltration (UF) and acid and antiscalant addition to the RO feedwater.

$MFI_{0.45}$ measurements for the same time period are given in Table 7.4. In this case the $MFI_{0.45}$ ($4.4s/l^2$) was quite similar to that found for the WRK-I water ($4.9-3.4s/l^2$). However, the $MFI_{0.45}$ value for the WRK-III feedwater was found to be 1900 times lower than the corresponding MFI-UF which indicates most particles in the WRK-III water are less than "0.45μm".

Table 7.4: $MFI_{0.45}$ determined at the IJssel Lake RO pilot plant for the influent WRK-III water and after ultrafiltration (UF) and acid and antiscalant addition to the RO feedwater.

Feedwater	$MFI_{0.45}$ (s/L^2)
WRK III	4.4
after UF	0.5
after acid and antiscalant addition	0.5

The particulate fouling potential was reduced significantly by the ultrafiltration step as the average MFI-UF value of the two measurements decreased to $1550s/L^2$ (Figure 7.2) and for the $MFI_{0.45}$ to $0.5s/L^2$. The MFI-UF and $MFI_{0.45}$ removal efficiencies for this step are given in Table 7.5. For the $MFI_{0.45}$ test, 90% of particles "$\geq 0.45\mu m$" were removed from the WRK-III feedwater by UF. As expected the removal efficiency for particles "$\geq 13\,000$ Da" (or "$\geq 9nm$") measured by the MFI-UF test was lower at $\approx 80\%$.

Table 7.5:   The removal efficiency per pretreatment step and overall treatment removal efficiency (%) at the IJssel Lake RO Pilot Plant as assessed by the MFI-UF and $MFI_{0.45}$ tests (- not measured).

Feedwater (sampling date)	Removal Efficiency of Pretreatment Step (%)		
	MFI-UF (s/L^2)		MFI$_{0.45}$ (s/L^2)
	1st Measurement	2nd Measurement	
after UF (December 2nd and 4th 1998)	80	80	90
after acid addition (November 27th)	-40	-	-
after acid and antiscalant addition (November 30th)	20	-	0
Overall treatment removal efficiency WRK -III to RO feedwater	80	≈ 80	90

The removal efficiency of UF was lower than expected which can be explained by particle counts measured during the same time period, presented in Figure 7.3. The number of particles in the WRK-III feedwater was very high and could not be determined in this research. Although, it was estimated previously by dilution to be in the range of 6 million per mL [5]. As observed in Figure 7.3 the *total* number of particles in the UF filtrate was ≈ 2600 per mL (sum of all channels in Figure 7.3). However, this value is considered too high as the number of particles in the UF filtrate for these membranes in full scale operation at PWN was found to be 100-500 per mL [19]. Furthermore, the pore size of the 150-200 000 Da ultrafiltration membrane was measured to be 12-22nm and a log removal of 4.7 was found in a MS2 virus challenge test [20]. Therefore, the lower removal efficiency is most likely due to leakages in the O ring connectors of the UF membranes and bacterial regrowth on the back of the membrane surface at the time of the measurements. Both conditions may have contributed to the relatively high MFI-UF measured for the UF filtrate. Thus, the lower removal efficiency of the UF step is not considered representative of normal operating conditions.

It should be noted that particle counters can be extremely sensitive to artifacts, arising from for example, noise or vibration, during the measurement. A further complication may have arisen with transparent bacterial particles in the size range of $>0.2\mu m$ which may have been falsely recorded by the counter as having a lower particle size of $0.05-0.10\mu m$ causing artificially high particles counts for this channel range (refer Figure 7.3).

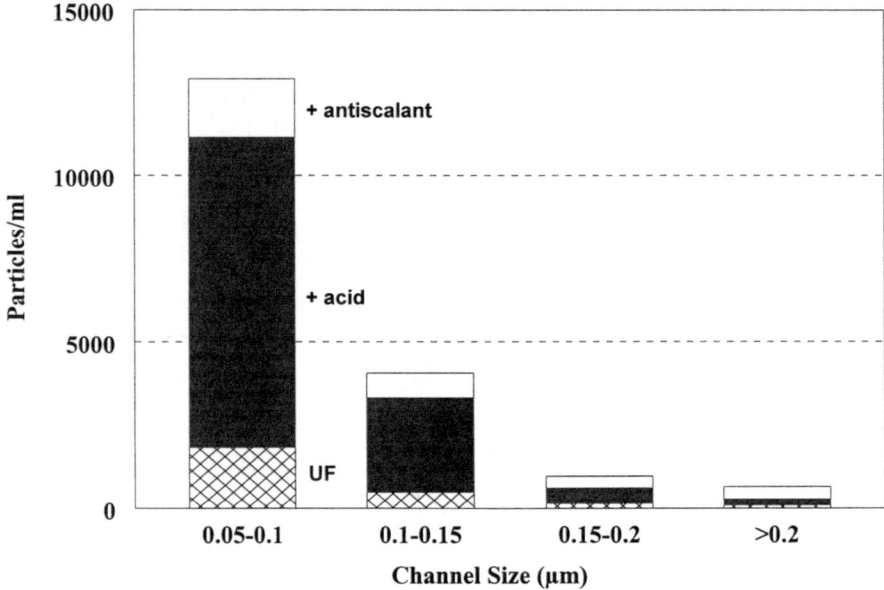

Figure 7.3: Particle counts in the RO feedwater after ultrafiltration at the IJssel Lake pilot plant and the increase in particle counts after acid and antiscalant addition to the RO feedwater.

After ultrafiltration, acid and antiscalant are added to the feedwater prior to reverse osmosis. From Table 7.5 the measured MFI-UF increased by 40% with the addition of acid. This suggests particles had been added along with the acid addition and as no increase was found for the $MFI_{0.45}$, the added particles were less than "0.45µm". Acid addition may also have decomposed particles into smaller sizes resulting in a higher cake resistance in accordance with Carmen Kozeny (Equation 7.4) and thus given a higher MFI-UF value. Inspection of Figure 7.3 confirms that acid addition was accompanied by a significant increase in the average particle counts measured to a *total* of 15 400 per mL, mostly in the 0.05-0.10 µm (increase of 9300 per mL) channel range. The increase in the MFI-UF due to these additional particles can be calculated via Equation 7.4. Smaller particles will have the most effect as the specific resistance of the cake measured on the MFI-UF membrane is indirectly proportional to the square of particle diameter. Assuming a cake porosity of 0.4, the MFI-UF was calculated to increase by only 5-10 s/L² for the additional increase in particles corresponding to the smallest particle size range of 0.05-0.10µm, respectively. However, the MFI-UF increased by 520s/L². This suggests that particles introduced directly by acid addition or by acid degradation (and which were retained by the MFI-UF membrane) were smaller than the smallest channel range of the particle counter i.e. ≤0.05µm.

Antiscalant addition also added particles to the RO feedwater, but to a lesser extent than with acid addition, the *total* number of particle counts increased to 18 600 per mL as seen in Figure 7.3. However, in this case the MFI-UF decreased by 20% after antiscalant addition which may be due to the bridging of smaller particles by antiscalant which increases the particle size and decreases the MFI-UF. The measurements of the RO feedwater after acid and after acid and antiscalant addition were not carried out on the same day. However, further support that this assertion is correct, is observed in Figure 7.4 which shows the MFI-UF of the RO feedwater after acid and after acid and antiscalant addition measured during November 1998. Despite the variability in the quality of the RO feedwater over time, the MFI-UF after only acid addition appears to be significantly higher than the trend for the MFI-UF after both acid and antiscalant addition over time.

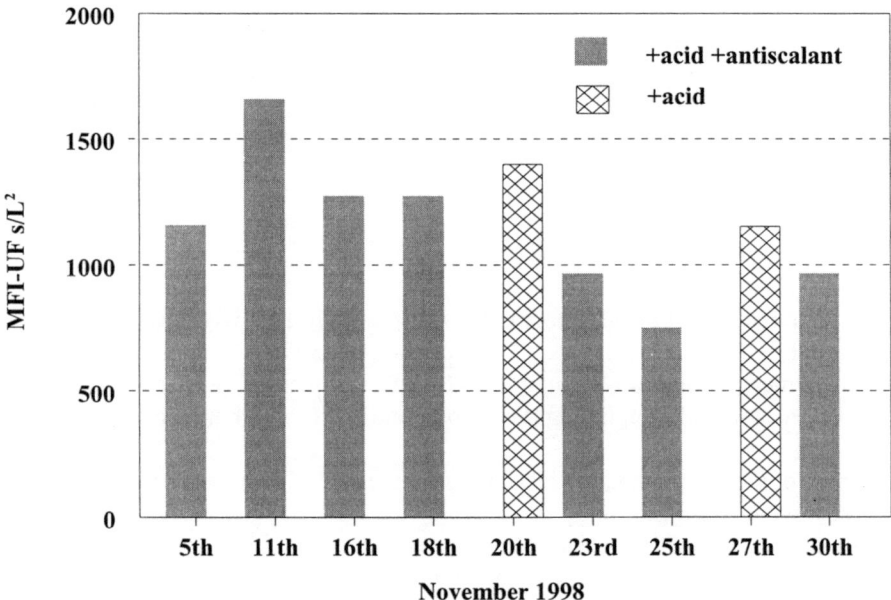

Figure 7.4: Quality of RO feedwater with acid and with acid and antiscalant addition at the IJssel Lake RO pilot plant in November 1998 as measured by the MFI-UF.

7.4.1.3 Capture of Particles in the MFI-UF test

The proportion of particles retained at the beginning and at the end of a MFI-UF test was examined using particle counts when applying the RO feedwater at the IJssel Lake plant after both acid and antiscalant addition. The results are presented in Figure 7.5. The test duration was circa 48 hours and carried out in December when the particle counts in the RO feedwater were markedly lower than when tested earlier (refer Figure 7.3). The variability of the RO feedwater was checked by particle counts at the beginning (t = 1 hour) and at the end of the test (t = 48 hours) and found to vary between the range of ≈10-40% for the smallest to largest channel size, respectively. Therefore, the percentage of particles retained by the MFI-UF test, reported in Table 7.6, were calculated for particle counts measured for the RO feedwater on that day.

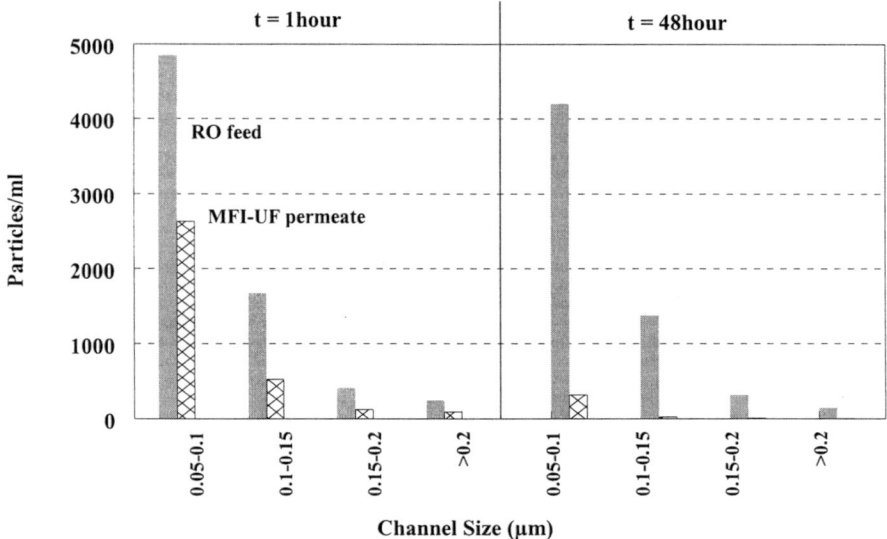

Figure 7.5: Comparison of particle counts for the RO feedwater (with acid and antiscalant addition) at the IJssel Lake RO pilot plant before (RO feed) and after (MFI-UF permeate) filtration through the MFI-UF test membrane. Particle counts were measured at the beginning (t = 1 hour) and end (t = 48 hours) of the experiment.

At the beginning of the experiment, particle counts for the MFI-UF permeate showed that some of the particles were passing through the pores of the MFI-UF membrane. Particles in the smallest channel size (0.05-0.1μm) showed the lowest retention of 50% (refer Table 7.6). This corresponds to early filtration time when blocking filtration is still occurring and to the initial stages of cake formation in the MFI-UF test. In Chapter 5 the pores of the polyacrylonitrile membrane were estimated by field emission scanning electron microscopy to have on average pores with length and breadth dimensions of 0.021 and 0.009μm, respectively and an average pore area of 150nm². Therefore, particles in the size range of 0.05 - 0.1μm should be retained. However, particles larger than the membrane pore may also pass through the pore. In literature, large flexible particles which have no interaction with the walls of a pore have been shown to pass a membrane [21 -22]. Thus, in the initial stages of filtration the passage of particles depends not only on size but also their charge, flexibility and configuration [21,22].

On the other hand, as mentioned in section 7.4.1.1, particles smaller than the membrane pore size may be retained by partial pore blocking during blocking filtration and as filtration proceeds and a cake has formed by particle bridging and entrapment within the cake itself. Thus, smaller particles are expected to be retained and measured in the MFI-UF (as the MFI test is based on cake filtration). This effect was confirmed by the increase to 90-≈100% of particles retained by the MFI-UF test membrane after 48 hours of filtration time for the smallest to largest channel range, respectively. Particles retained in the MFI-UF test where the pore length and/or breadth dimension acted as the limiting dimension preventing the passage of a particle could not be examined, as these particles will be smaller than the smallest channel size of the particle counter

($\geq 0.05\mu m$). However, most likely a proportion of these particles (as observed after acid addition to the RO feedwater in section 7.4.1.2) are also retained and increasingly so over time. Similarly, the $MFI_{0.45}$ and the $MFI_{0.05}$ membranes will retain particles smaller than $0.45\mu m$ and $0.05\mu m$, respectively, by the mechanisms described above.

Table 7.6:   Variability in particle counts for the IJssel Lake RO feedwater containing acid and antiscalant during the MFI-UF experiment and the per centage of particles retained by the MFI-UF membrane from the feedwater at the beginning (t = 1 hour) and end of the experiment (t = 48 hours).

Channel Size (μm)	Variability in RO feedwater particle counts (%) during experiment	Particles initially retained in MFI-UF test (%)	Particles retained in MFI-UF test at 48 hours (%)
0.05-0.10	10	50	90
0.10-0.15	20	70	≈100
0.15-0.20	20	70	≈100
>0.20	40	60	≈100

These results demonstrate that the MFI-UF test captures and measures smaller colloidal particles than other commonly used methods for characterising the particulate fouling potential of a feedwater e.g. the $MFI_{0.45}$ and SDI. Moreover, these results show that the variable removal efficiencies for a given pretreatment step found by the SDI may be due to the fact that one test measurement was conducted during blocking filtration and another conducted during cake filtration, leading to variable particle retention and thus variable SDI results. This highlights the importance of the theoretical basis i.e. cake filtration of the $MFI_{0.45}$, $MFI_{0.05}$ and MFI-UF tests. Thus, depending on which membrane filtration process is to be employed e.g. ultrafiltration or reverse osmosis either the $MFI_{0.45}$, $MFI_{0.05}$ or the MFI-UF test is recommended to characterise the particulate fouling potential of a feedwater. Similarly, to assess the efficiency of a pretreatment process for the removal of selected particles, the $MFI_{0.45}$, $MFI_{0.05}$ or the MFI-UF test is recommended alone or in combination.

7.4.2   MFI-UF for Flux Decline Prediction

7.4.2.1   Simplified Flux Decline Prediction

Membrane cleaning is commonly recommended when a 15-20% decrease in the normalised flux or increase in pressure drop of an installation is observed. The MFI-UF measured for the RO feedwater at the IJssel Lake and River Rhine pilot plants were used to predict the time for a 15% decline in permeate flux to occur applying the simplifying assumptions that the particle deposition factor (Ω) and the cake ratio factor (ψ) were 1 (Equation 7.16). In the first scenario the pressure drop across the reverse osmosis membrane and fouling cake layer was taken to be 10 bar and the MFI-UF was corrected to 10 bar by applying the global compressibility coefficient. (However, it should be noted that pressure compression applying a compressibility coefficient was only verified to 2 bar in Chapter 6, and therefore, extrapolating to this high value may reduce the accuracy of the model). For this scenario the average predicted minimum (as Ω = 0) run time for the IJssel Lake RO pilot plants was 9.5 days at 80% recovery (refer Table 7.7). Whereas, the average predicted minimum run times at the River Rhine plant were more than 10 times longer, 99 days, at 85% recovery. Much longer run times were expected at the River Rhine pilot plant due to the better quality of the RO feedwater after pretreatment (refer section 7.4.1.1) where the MFI-UF measured was close to the detection limit of the test.

Table 7.7:   Minimum run time for a 15% decline in permeate flux to occur at the IJssel Lake and River Rhine RO pilot plants calculated from MFI-UF measurements of the RO feedwater corrected to a pressure drop (ΔP_r) in the RO system of 10 and 2 bar.

Minimum Run Time for 15% flux decline in the RO pilot plant (days)					
IJssel Lake			River Rhine		
Date (1998)	$\Delta P_r = 10$ bar	$\Delta P_r = 2$ bar	Date	$\Delta P_r = 10$ bar	$\Delta P_r = 2$ bar
5th Nov.	9	31	Sept. 1998	88	294
11th Nov.	6	22	Feb. 1999	122	405
16th Nov.	8	28	March 1999	78	260
18th Nov.	8	28	Apr. 1999	109	364
20th Nov. (a)	8	28			
23rd Nov.	11	37			
25th Nov.	14	48			
27th Nov. (a)	9	31			
30th Nov.	11	37			

(a) single stage system

These results, based on MFI-UF measurements, include fouling for particles "≥13 000 Da". When the same calculations were made for fouling due to larger particles i.e. "≥0.45 μm" using $MFI_{0.45}$ measurements for the RO feedwater, the predicted minimum run times were dramatically longer, as cakes composed of larger particles have a lower specific resistance (Equation 7.4). In the case of the IJssel Lake plant a 15% flux decline was predicted to occur in 99 years. Similarly, for the River Rhine plant this figure was 99-290 years for the February and August measurement, respectively. However, minimum run time predictions based on $MFI_{0.45}$ measurements are unrealistically long as pilot plant experience at AWS has shown that $MFI_{0.45}$ values between 0.4-0.6 gives a RO plant running time of three to eight months [16]. Therefore, the general correlation observed at the pilot plant most likely includes fouling due to smaller colloidal particles although not measured in the $MFI_{0.45}$ test.

A comparison of the observed minimum run times at the River Rhine plant to that predicted by the MFI model proved difficult to make as the RO system was cleaned due to other causes during the period that the MFI-UF measurements were taken. In addition, the MFI-UF measurements of the RO feedwater, were made prior to antiscalant addition which based on the results found in section 7.4.1.2 for the IJssel Lake RO feedwater, increased the fouling potential of the feedwater.

In the case of the PWN two stage systems, the HA and HC plants had not been cleaned since the membranes were installed 17 and 7 months ago, respectively, as no fouling was evident. Likewise the HT had not been cleaned since January 1998 when it was cleaned for preventative purposes and no fouling was observed 6 months later when the MFI-UF measurements were carried out. However, the predicted run times at the IJssel Lake plant were based on measurements taken during November and as discussed earlier in section 7.4.1.2 this coincided with problems in the UF installation which resulted in a lower UF efficiency than expected. Therefore, the predicted minimum time run time are expected to be only relevant for this period. Nevertheless, no fouling was observed during November and early December and thus, in practice the minimum run times are longer than those predicted by the MFI model.

The shorter run times may be explained in part by the fact that the pressure drop of 10 bar is principally to overcome the high resistance of reverse osmosis membranes. The actual pressure drop over the fouling cake layer is most likely only ≈ 15-20% (1.5-2 bar) of this value. Therefore, the same calculation was made as above with pressure correction of the MFI-UF to 2 bar instead of 10 bar. (Note, this pressure drop in now within the verified pressure range for correction of the MFI-UF value for cake compression effects). Predicted minimum run times for this pressure drop were extended by a factor of 3.3 for both pilot plants and are now more realistic for the River Rhine plant. However, they remain shorter than that observed at the IJssel Lake plant.

Assuming that all the smaller particles causing fouling are retained by the MFI-UF membrane (" ≥ 9nm" refer section 7.4.1.3), the differences between the run times predicted and that observed may be explained by the following reasons; (i) the particle deposition factor is less than 1 i.e. not all particles are deposited on the membrane and (ii) the cakes formed on the MFI-UF membrane are dissimilar to that formed on the reverse osmosis membrane i.e. the cake ratio factor is not 1 and (iii) the MFI-UF is carried out under constant pressure filtration, thus, the flux is decreasing during the test. Whereas, most RO installations operate under constant flux with the pressure increasing over time to maintain flux. Thus, the pressure is increasing over the cake formed on a RO membrane and the choice of which pressure to apply in the flux decline prediction model is questionable. It may then be more appropriate to determine the MFI-UF in constant flux mode to simulate the cake formed at the membrane and extrapolate the MFI-UF value to the flux of the installation to predict fouling. These three factors are examined in the following sections.

7.4.2.2   The Particle Deposition Factor

The particle deposition factor was first determined at the IJssel Lake RO pilot plants and the results are presented in Table 7.8. Surprisingly, the deposition factor was negative for both the single and the two stage systems. The deposition factors were not significantly different $\approx$-1.12

for the three RO pilot plants at 80% recovery and was halved in the single stage system, -0.48 as expected from the concentration factor at this recovery.

A positive deposition factor indicates particles are being deposited as they pass through the system while a negative factor indicates the number of particles in the concentrate exceeds the incoming flux (taking into account the concentration factor). The two stage RO systems have been in operation for 6 - 17 months without cleaning. From literature the formation of a thin cake under cross flow conditions was expected with prolonged operation and the removal of particles from the cake may have contributed to the higher MFI-UF measured for the concentrate [9].

Table 7.8:   The particle deposition factor calculated at the IJssel Lake RO pilot plants.

Date 1998	Recovery (%)	MFI-UF feedwater (s/L^2)	MFI-UF concentrate (s/L^2)	Ω factor (-)
23rd Nov. (HT)	80	1600	15 100	-1.11
25th Nov. (HA)	80	1300	12 400	-1.13
30th Nov. (HC)	79	1600	14 400	-1.12
27th Nov. [a] (HD)	52	1900	5 100	-0.55

(a) single stage

To confirm negative particle deposition factors, the particle deposition factor was also determined at the River Rhine RO pilot plant. In this case the particle deposition factor was measured at different pressures varying from 0.5-2 bar. The MFI-UF of the feedwater and concentrate were corrected to 2 bar using the respective *individual* compressibility coefficients determined, $\omega = 1.35$ and 1.08. (It should be noted the use of the individual compressibility coefficient is preferable for the River Rhine RO plant as the global coefficient effectively cancels in Equation 7.14. Whereas, at the IJssel Lake plant all the determinations were at an applied transmembrane pressure of 1 bar and can therefore be compared). The particle deposition factors calculated at the River Rhine RO pilot plant, reported in Table 7.9, were also found to be negative. The particle deposition factor ranged from -0.44 to -0.58 over the sampled time period. This variation is expected to be due to minor fluctuations in RO feedwater quality over time.

The negative particle deposition factors obtained for both the River Rhine and IJssel Lake plants precludes their use in the flux decline prediction model employed in section 7.4.2.1. Nevertheless the negative particle deposition factors indicate that little or no flux decline should be observed for the pilot plants.

The negative deposition factors may be as a result of changes in the composition of the cake formed on the RO membranes over time. The MFI-UF (and $MFI_{0.45}$) are carried out in dead end mode while the RO pilot plants are operated in cross flow mode. Particles formed in a cake in cross flow mode are reported to change over time due to the forces acting on the particle in tangential flow [23]. Convection and drag forces transport particles to the membrane surface while shear induced diffusion, Brownian diffusion and inertial lift may cause back transport of colloidal particles from the membrane to the bulk solution [14,24]. The net result of these opposing forces depends strongly on the size of the particle [25]. Consequently, selective particle deposition occurs, changing the cake particle size distribution and resistance. In general, a cake of higher resistance is observed due to the more favourable deposition of smaller particles [23]. The lateral migration of the larger particles across streamlines has been explained in literature by two mechanisms. Belfort and co-workers showed inertial lift (commonly referred to as the "tubular pinch effect") is important for large particles whereby, larger filtrated particles migrate away from the membrane wall reaching an equilibrium at some eccentric position in the flow [13]. The other mechanism proposed is shear induced diffusion, whereby, particles moving in different streamlines tumble over each other in shear flow and their random interaction "nudges" particles in the direction of lower particle concentration [23,26,27].

Table 7.9: The particle deposition factor calculated at the River Rhine RO pilot plant (85% recovery) from the MFI-UF measured for the feedwater and concentrate at 0.5-2.0 ΔP and corrected to a pressure drop in the RO system of 2 bar using the individual compressibility factors of the feedwater ($\omega = 1.35$) and concentrate ($\omega = 1.08$).

Date 1999	Applied MFI-UF test pressure (ΔP bar)	MFI-UF feedwater (s/L^2)	MFI-UF concentrate (s/L^2)	Ω factor (-)
29th March	1.50	200	2000	-0.58
31st March	2.00	200	2000	-0.58
1st April	0.54	200	1800	-0.41
6th April	1.03	250	2300	-0.44

Shear induced and/or inertial lift particle classification of the cake formed on the membranes at the RO pilot plants, may result in the cake being composed predominantly of smaller particles. This would result in cakes on the RO membrane with a higher resistance than those found in dead end filtration. Indeed, Chellam [23] demonstrated that cakes formed in cross flow have a lower permeability than that formed from dead end filtration. If this was the case at the RO pilot plants, the remaining larger particles in the bulk concentrate would result in a lower MFI-UF measured giving a positive deposition factor. However, most likely the cake and the particle deposition factor are not static. A proportion of the smaller particles in the cake might be removed from the top layer over time resulting in higher MFI-UF values, as was found for the concentrate at the RO pilot plants and hence result in a negative deposition factor.

As suggested by Chellam, changes in cake composition and specific cake resistance in cross flow mode may partially explain the poor correlation between the $MFI_{0.45}$ and membrane fouling rates [23]. To correct for changes in cake composition and improve fouling prediction using MFI-UF measurements, the particle deposition factor could be determined of mono dispersed colloids of various sizes in cross flow filtration. Alternatively, the particle deposition factor could be determined using the $MFI_{0.45}$, $MFI_{0.05}$ and MFI-UF tests for the retention of particles of selected sizes. The particle deposition factor should then be determined after initial start up and at various times during operation to examine the behaviour of the particle deposition factor over time. Initially, the particle deposition factor might be expected to be positive as particles are deposited and a cake builds up. When the cake has reached a certain height the particle deposition factor may vary depending on the size of the particle as to whether the cake is at equilibrium ($\Omega = 0$), particles are removed by the tangential flow across the membrane surface ($\Omega < 0$) or particles are deposited ($\Omega > 0$).

7.4.2.3   The Cake Ratio Factor

In section 7.4.2.1 flux decline prediction was calculated for the RO pilot plants assuming the cake formed on the MFI-UF membranes was identical to that formed on the RO membranes (i.e. the cake ratio factor is 1). However, the cake formed on the target membrane will be the result of the flow mode, the prevailing flux and pressure conditions at the membrane surface and the increasing salinity of the feedwater with recovery. Differences in the flow mode may be accounted for by further investigation of the particle deposition factor, similarly differences in flux and pressure may be improved by the determination of the MFI-UF in the constant flux mode. On the other hand the effect of increasing ionic strength on the MFI-UF is unknown.

Increasing salinity (ionic strength) of a feedwater has been reported to exacerbate fouling in micro, ultrafiltration and reverse osmosis systems by; reducing both cake porosity and the "effective" particle size (which increase the specific cake resistance), increasing the particle deposition rate in cross flow and adhesion forces of particles to the membrane surface [15, 28-30]. These effects have been explained by changes in the diffuse double layer surrounding both the particle and membrane. Typically, small particles present in surface water are negatively charged and stable due to their high zeta potential (a measurement of the electrokinetic potential of the surface) [15]. Similarly, the membrane surface and pore also often possess a negative charge and develop a double layer in water, a polar medium [15]. An increase in ionic strength compresses the double layer around the particle and membrane surface which results in the aforementioned effects which includes an increase in the specific cake resistance.

A preliminary experiment to assess the effect of increasing salinity on the specific cake resistance measured in the MFI-UF index was carried out with tap water with the addition of sodium chloride to an ionic strength up to 0.2 M. The MFI-UF, presented in Figure 7.6 as a function of ionic strength, initially increased with increasing ionic strength reaching a maximum at 0.1M. Thereafter, the effect was found to diminish and the MFI-UF decreased to only 13% above that measured at 0.01M. A similar trend was observed by Bacchin et al [31] for the specific cake resistance of bentonite in cross flow ultrafiltration with increasing addition of KCl in the range of 1×10^{-5} to 0.1M. The bentonite specific resistance was independent of ionic strength up to 0.001M, after which it increased to a maximum at ≈ 0.003M and subsequently decreased to *below* that measured at 1×10^{-5}M. Bacchin et al [31] demonstrated the initial increase in specific cake resistance from 0.001 to 0.003M was due to a reduction in cake porosity caused by a decrease in the interparticle distance between particles in the cake. While the subsequent decrease in specific resistance was due to the increase in ionic strength above the critical concentration of coagulation (CCC) at which the zeta potential reduced to zero. Hence, particles aggregated into larger sizes leading to a more permeable cake via the Carmen-Kozeny equation (Equation 7.4).

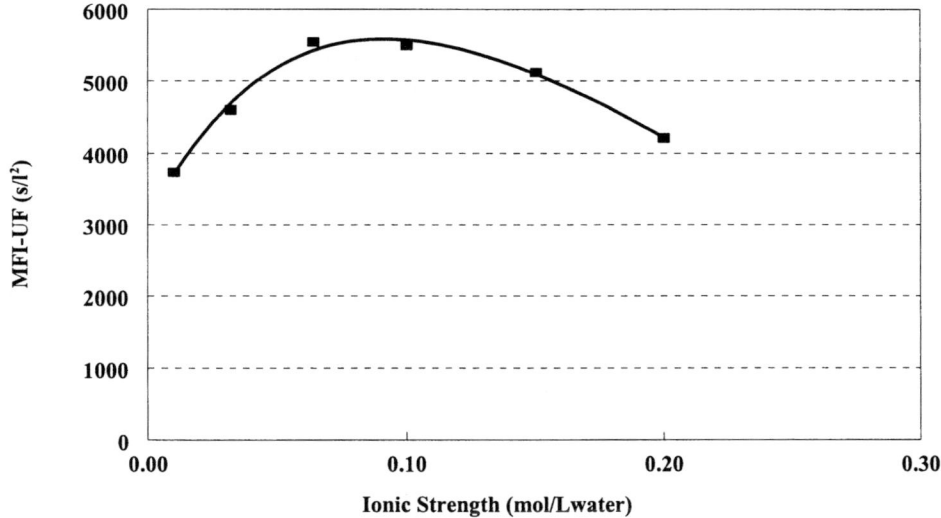

Figure 7.6: The MFI-UF of tap water as a function of ionic strength (via addition of sodium chloride).

The maximum MFI-UF with tap water occurred at 0.1M, which is a factor 33 times greater than that observed by Bacchin et al for suspensions of bentonite. The maximum observed with ionic strength will depend on the critical concentration of coagulation of the particles present in a feedwater, some particles may require a higher electrolyte concentration to induce coagulation. In addition, natural organic matter (NOM) e.g. humic acids present in surface water may contribute to the location and magnitude of the maximum. Gosh and Schnitzer [32] established two hypothetical configurations of NOM. At low ionic strength, high pH and low solute concentration NOM has been characterised as flexible linear colloids and in response to increases in ionic strength and solute concentration and decreasing pH, NOM macromolecules coils. Thus, at the moderate pH of surface water, NOM may increasingly be present as rigid compact sphero colloidal molecules with increasing recovery and concomitant increase in salinity and solute concentration [33].

From the discussion above and the effect demonstrated in Figure 7.6 of increasing ionic strength on the MFI-UF index, an ionic strength correction of the MFI-UF is most likely required. The ionic strength of the RO feedwater at the two pilot plants is 0.01M and increases up to 0.05 and 0.07M at 80 and 85% recovery, respectively. To correct for increasing salinity in the pilot plant concentrates a function was fitted to the MFI-UF measured for tap water as a function of ionic strength. Applying this function, the MFI-UF (of tap water) was calculated to increase by 12%, 28% and 31% at the ionic strength corresponding to the IJssel Lake RO pilot plants at 52, and 80% recovery and the River Rhine plant at 85%. Assuming this factor could be applied to the RO feedwater, the MFI-UF measured for the RO feedwater was corrected by this factor for the system recovery and used to recalculate the particle deposition factor (refer Table 7.10).

Table 7.10: Particle deposition factor (Ω) calculated with and without ionic strength correction at the IJssel Lake and River Rhine RO pilot plant.

Date 1998	IJssel Lake		River Rhine	
	Ω factor	Ω factor (corrected for ionic strength)	Ω factor (-)	Ω factor (corrected for ionic strength)
29th March			-0.58	0.00
31st March			-0.58	0.00
1st April			-0.41	0.12
6th April			-0.44	-0.10
23rd Nov. (HT)	-1.11	-0.56		
25th Nov. (HA)	-1.13	-0.66		
30th Nov. (HC)	-1.12	-0.56		
27th Nov. (HC) [a]	-0.55	-0.22		

(a) single stage

The deposition factor for the River Rhine plant was then found to be positive and close to zero in 75% of the cases. Although, the deposition factor in the IJssel lake plant has been halved it remains negative. This may be due to the differences in cake compressibility between the feedwater and concentrate with ionic strength. For the IJssel lake plant the global compressibility coefficient was applied to both feedwater and concentrate and this approach is inappropriate in calculating the particle deposition factor. The compressibility coefficient found in the River Rhine 85% concentrate (1.08) decreased in comparison to the feedwater (1.35) which may be due to a reduction in size of inorganic and organic particles at the higher ionic strength. Tadros and Mayes [34] observed a reduction in the compressibility coefficient for silica and polystyrene particles with increasing electrolyte concentration. Although, smaller particles may be more resistant to compression, at low ionic strength double layer repulsion forces between particles are quite high and thus the particle may undergo particle rearrangement in the cake more easily causing a decrease in cake porosity with increased applied pressure giving a higher compressibility coefficient [34]. Whereas, particle rearrangement may be more restricted at higher ionic strengths due to the reduction in the double layer and a lower compressibility coefficient may then be obtained. Therefore, the individual compressibility coefficients of the feedwater and concentrate at the IJssel Lake plant need to be determined in subsequent research to calculate the particle deposition factor.

A correction for ionic strength is also necessary in the flux decline prediction model and for characterising the fouling potential of a feedwater. However, in order to determine a general correction method for ionic strength, more feedwater types need to be investigated of varying inorganic and organic composition as the critical concentration of coagulation is expected to be dependent on feedwater composition as discussed above. Moreover, the effect of increasing ionic strength on the MFI-UF test membrane needs to be determined. Increasing ionic strength has been reported to decrease rejection despite a reduction in membrane permeability [33,35].

7.4.3   MFI-UF in Constant Flux Mode

7.4.3.1   Determination of the Fouling Index I at Constant Flux

As suggested in section 7.4.2.1 more accurate flux decline prediction may be obtained by determining the MFI-UF in constant flux mode. Results of a preliminary experiment of the MFI-UF operated under constant flux (CF) mode for a low fouling (tap water) and a high fouling feedwater (10% canal water dilution) are presented as the fouling index I over time in Figure 7.7. In constant flux filtration the pressure increases over time to maintain the selected constant applied flux, in this case, 90 l/m²hr. The development of the pressure increase over time for the two feedwater is included in Figure 7.7.

The same filtration mechanisms which may occur under constant pressure filtration were expected to occur in the constant flux mode. Thus, the initial sharp increase in slope in the pressure vs time plot was expected to be caused by blocking filtration which resulted in the initial sharp decrease in the fouling index (deviations in the curve are a consequence of the manual adjustment of pressure). The onset of cake filtration for both feedwater was observed by linearity in the pressure versus time plot as was expected from Equation 7.8. This was evident in the I versus time plot as a stable or minimum fouling index I and occurred for both feedwater after ≈ 1.5 hours of filtration. Thus, the I and hence the MFI-UF could be determined within a few hours of filtration, in marked contrast to when the MFI-UF is conducted under constant pressure mode, where the MFI-UF could take up to 20 hours to stabilise for tap water filtration (refer Chapter 5 section 5.4.1).

The MFI-UF calculated from the I index was two times higher for the more fouling diluted canal water (5600 s/L²) than for the tap water (2400 s/L²) as expected. Furthermore, the subsequent increase in the I index due to cake clogging and/or cake compression occurred at an earlier time and was more severe for the diluted canal water than for the tap water. This can be attributed to the higher concentration of particles present in the diluted canal water.

The above results demonstrate that the MFI-UF could be determined in the constant flux mode and after a shorter time than in the constant pressure mode. The most noticeable difference between the behaviour of the I index (or MFI-UF) over time for the two filtration modes is the duration of the stable region. In constant pressure mode the MFI-UF was found to be stable for 70 hours with prolonged filtration of tap water (refer Figure 5.2 Chapter 5). Whereas, in the constant flux mode, the duration of the stable period was much shorter and depended on the fouling potential of the feedwater. Stability was limited in the case of the 10% canal water dilution, occurring for only a period of 50 minutes and was observed in the I index versus time plot as a minimum. Whereas, for tap water the I index was stable for 1.5 hours and the minimum was not as distinct as for the diluted canal water.

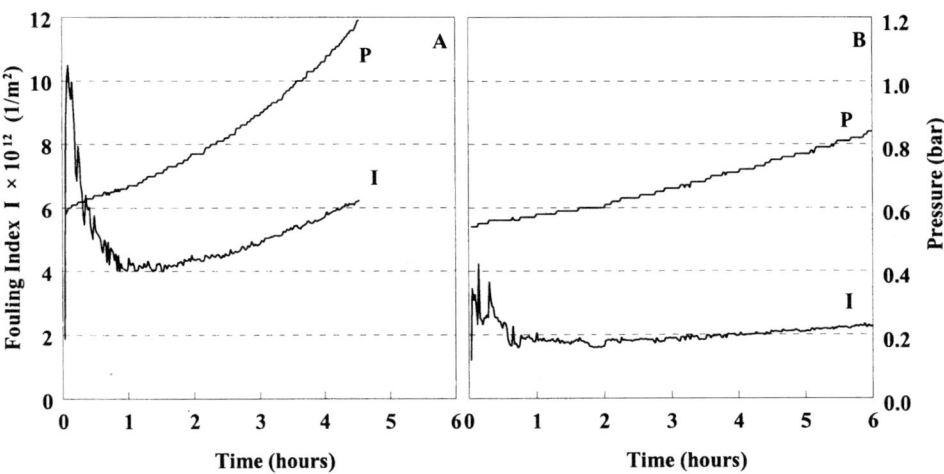

Figure 7.7: The fouling index I and pressure development P over time for (A) a 10% canal water dilution and (B) tap water determined at a constant applied flux of 90 l/m²hr. (The fouling index I is read from the left Y-axis and pressure P from the right Y-axis).

7.4.3.2   Linearity of Fouling Index I with Colloidal Concentration (CF mode)

In Chapter 6 (section 6.4.1.2) a linear relationship was demonstrated between the MFI-UF index and particulate concentration determined under constant pressure filtration. In previous research, the $MFI_{0.45}$ index was shown to be linear with particulate concentration when determined under constant flux filtration [37]. As linearity of the MFI-UF (or fouling index I) with particulate concentration is a consequence of cake filtration (via Equation 7.2) and cake filtration was shown to occur in the previous section for tap and 10% diluted canal water, linearity was also expected for the MFI-UF index in CF mode. However, to confirm this the fouling index I was determined of tap water (low fouling) and canal water (high fouling) dilutions under a constant applied flux of 75 l/m²hr, the results of which are presented in Figure 7.8 and 7.9, respectively.

For both feedwater the fouling index I decreased with increasing dilution. Cake filtration was evident as a stable or minimum fouling index I over time. As suggested in the previous section, the duration for which the fouling index I was stable before cake clogging and/or cake compression was observed, is dependent on particle concentration. Thus, for canal water dilutions the fouling index I was stable for circa 2 hours, 1 hour and 30 minutes for 9%, 18% and 35% dilutions, respectively (refer Figure 7.8). Whereas, the fouling index I was stable for the entire duration of the experiments for all the tap water dilutions with no cake clogging and/or cake compression observed.

The fouling index I determined for the canal and tap water dilutions is plotted as a function of dilution in Figure 7.10. Linearity was found for both feedwater, the regression coefficient calculated for canal and tap water were 0.977 and 0.960, respectively. These results demonstrate that the fouling index I is proportional with particle concentration and further supports that cake filtration occurs in the MFI-UF test when determined under constant flux filtration.

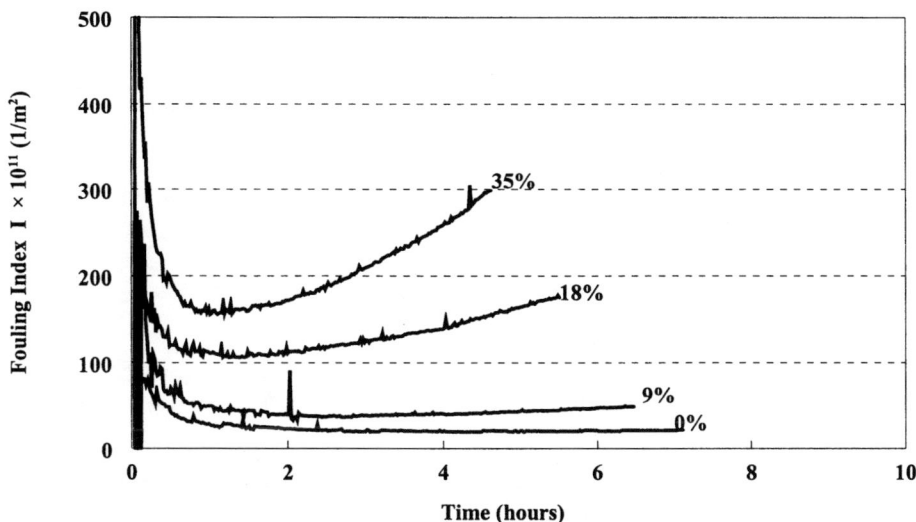

Figure 7.8: The fouling index I measured for canal water dilutions over time in the constant flux mode (75 l/m²hr).

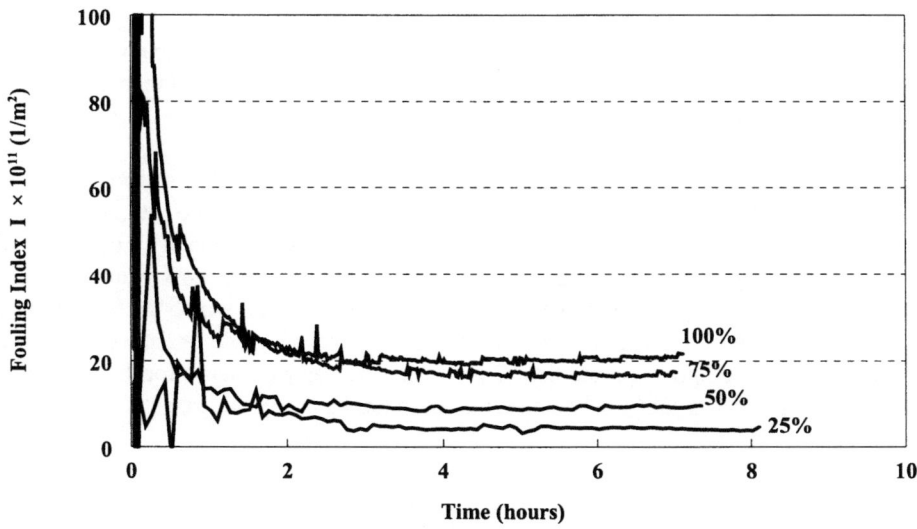

Figure 7.9: The fouling index I measured for tap water dilutions over time in the constant flux mode (75 l/m²hr).

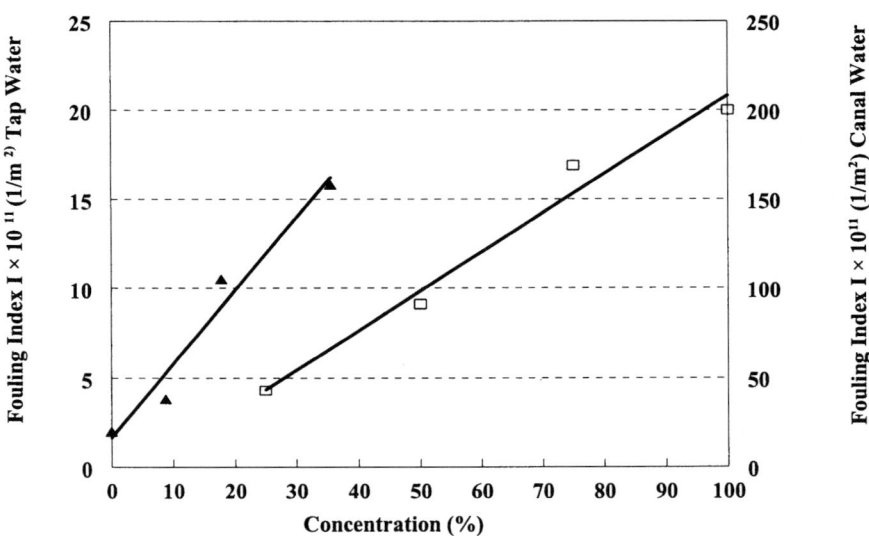

Figure 7.10: Relationship of the fouling index I *vs.* dilutions of canal water (▲)and tap water (□) determined in the constant flux mode (75 l/m²hr).

7.4.3.3   Effect of Applied Flux

The effect of applied flux in the range of 70 -≈110 l/m²hr on the fouling index I measured for the 10% canal water dilution and tap water in the constant flux mode was examined. To facilitate a comparison of the fouling index I at the different applied test fluxes, the fouling index I was plotted versus the filtered volume, presented in Figures 7.11 and 7.12 for the canal water dilution and tap water, respectively. (In Figure 7.7 the applied fluxes are the same and therefore there is no difference as to whether I is plotted versus filtered volume or time).

For the 10% canal water dilution, the fouling index I could be determined in all the tests after a filtered volume of 15-20 litres when a minimum I value occurred. When the filtered volume is translated to time, the fouling index I could be determined at an earlier time at a higher applied test flux e.g. at the two extremes of applied flux, 105 and 70 l/m²hr, the I value reached a minimum at 0.5 and 1.25 hours, respectively. However, the duration for which the fouling index I was stable, shortened with increasing applied flux, as cake clogging and/or cake compression was also exacerbated at a higher applied flux. This was particularly dramatic for the test carried out at the applied flux of 105 l/m²hr where the fouling index I sharply increased after only circa 30 litres of filtered volume.

In the case of tap water the fouling index I could be determined at a similar filtered volume or time in all the tests. The effect of applied flux on cake clogging and/or cake compression was not so pronounced with increasing applied flux (refer Figure 7.12) as with the canal water dilution. This is expected to be due to the lower particle concentration in the tap water which may have delayed the onset of cake clogging and/or cake compression. Nevertheless, results of the 10%

canal water dilution suggest that cake filtration may hold for only a specific period of time dependent on the applied flux.

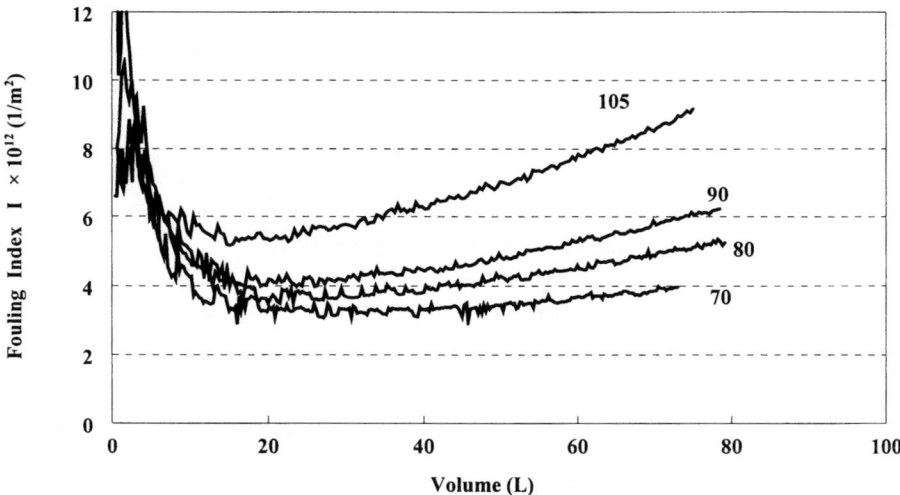

Figure 7.11: The effect of an increase in the applied test flux on the fouling index I measured for a 10% canal water dilution over time in the constant flux mode.

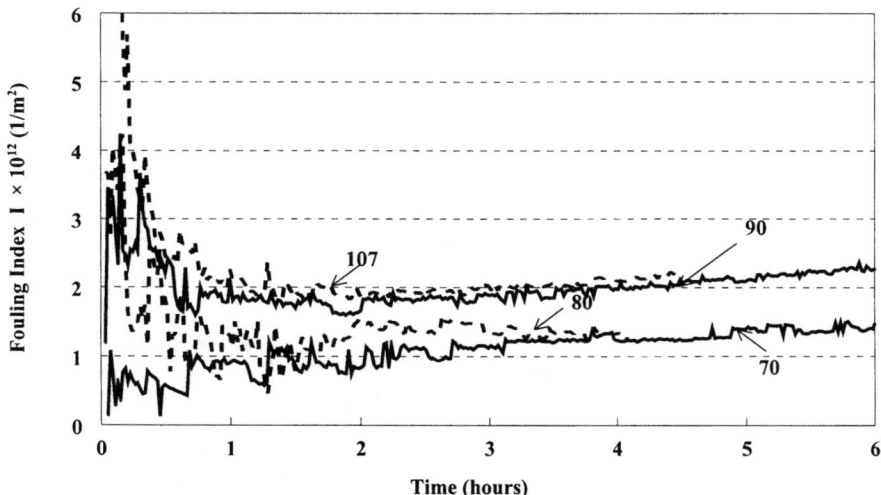

Figure 7.12: The effect of an increase in the applied test flux on the fouling index I measured for tap water over time in the constant flux mode.

For both feedwater an increase in applied flux resulted in an increase in the fouling index I. As in constant pressure filtration, cake compression may occur at a higher applied test flux. At a higher flux the particles impinge on the membrane surface at a higher velocity and the cake may then assume a lower porosity i.e. a denser cake with a higher specific resistance than at a lower applied flux or velocity [36]. Consequently, a higher fouling index I was found. For the 10% canal water dilution the effects of increased flux and pressure could be separated in three of the four tests, by comparing the fouling index I at the same pressure of 0.75 bar. The fouling index I determined at this pressure is reported in Table 7.11 and was found to increase by ≈9% for an increase in applied flux from 70 to 90 l/m²hr. A similar comparison could not be made for the tap water data, as no common pressure value occurred. However, a plot of the fouling index I determined for both feedwater increased linearly with applied flux for both feedwater suggesting that an increase in flux caused cake compression. Note that in this case the fouling index I presented in Figure 7.13 is also a function of the pressure as the fouling index I reached a minimum or a stable value at different pressure values in the tests.

Table 7.11: Effect of increased applied flux on the fouling index I in the 10% canal water dilution experiments compared at the same pressure of 0.75 bar.

Applied Test Flux (l/m²hr)	Fouling Index I (m⁻²)
70	4.00×10^{-12}
80	4.25×10^{-12}
90	4.34×10^{-12}

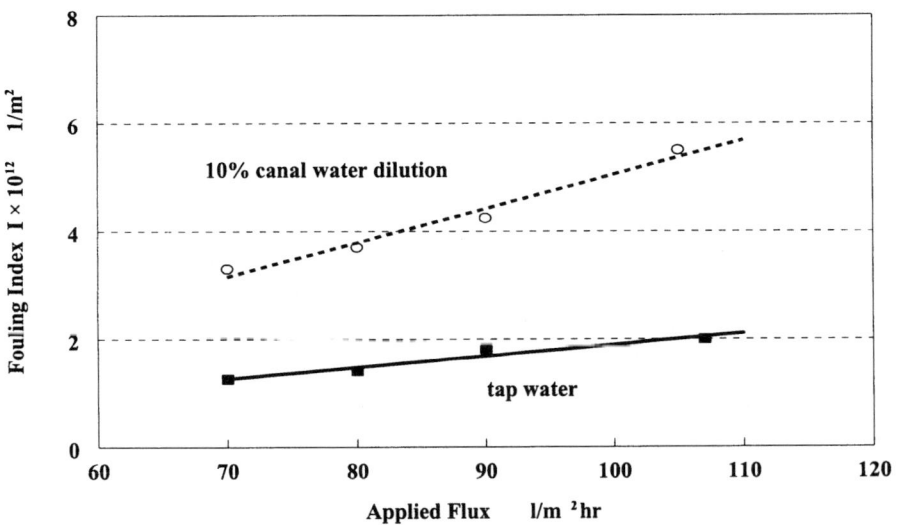

Figure 7.13: The effect of applied flux on the fouling index I determined for tap water and a 10% canal water dilution in constant flux mode filtration.

A comparison between the fouling index I determined for the two feedwater at a given flux in Figure 7.13, shows the fouling index I was more than two times higher for the more fouling feedwater, diluted canal water, than the lower fouling feedwater i.e. tap water as expected. Moreover, linear regression of the data for each feedwater demonstrated that the slope of the line determined for the diluted canal water was three times higher than that determined for tap water. This indicates that the cake formed with the 10% canal water dilution was more compressible than the tap water.

The MFI-UF index calculated from the I value obtained in constant flux mode for the diluted canal water and tap water, presented in Figure 7.13, are summarised in Table 7.12. As can be seen the relative fouling potential of the feedwater depends on which flux is chosen. For the MFI-UF developed under constant pressure filtration, the MFI-UF index determined at a non standard pressure can be satisfactorily corrected to the reference pressure of 2 bar so that the fouling potential of feedwater can be compared. Similarly, for the determination of the MFI-UF in constant flux mode a reference flux needs to identified e.g. 100 l/m²hr and a method developed to correct the MFI-UF to the reference flux when conducted at non reference conditions.

Table 7.12: The MFI-UF determined under constant flux mode for a 10% canal water dilution and tap water at various applied test fluxes.

	MFI-UF (s/l²)	
	10% canal water dilution	Tap water
70 l/m²hr	4300	1700
80 l/m²hr	4900	1800
90 l/m²hr	5600	2400
105 l/m²hr	7200	
107 l/m²hr		2600

Preliminary results of the MFI-UF determined under constant flux filtration are promising, whereby, the MFI-UF index can be determined within a short time period and will more likely simulate the fouling conditions at the surface of membranes e.g. RO operating at a constant flux. In addition, the fouling index I was demonstrated to be linear with particle concentration when determined in this mode. However, to model and predict fouling based on MFI-UF measurements determined in the constant flux mode, lower applied fluxes e.g. 20-30 l/m²hr for RO need to be investigated. The minimum flux that could be examined in this study was 70 l/m²hr due to the limitation of the flow meter combined with the higher flux of the MFI-UF test membrane. Therefore, the next steps in the development of the MFI-UF in constant flux mode is to build an automated MFI-UF system with a more accurate flow meter for lower flows. After which, feedwater from different sources and after various pretreatment steps need to be examined to identify a reference flux for the test and to develop a method to correct the MFI-UF for flux effects on the specific cake resistance.

7.5   Conclusions

The MFI-UF of the influent feedwater to the RO Pilot Plants was circa 700-1900 times higher than that of the respective $MFI_{0.45}$ due to the retention of smaller particles ("$\geq$13 000 kDa") in the MFI-UF test. Slow sand filtration was the most successful pretreatment process at the River Rhine Plant with consistently 90-$\approx$100% removal of particles in the size range measured by the MFI-UF and $MFI_{0.45}$ tests. Whereas, the SDI gave variable results for this step at 40->90%. The fouling potential of the RO feedwater at this plant after extended pretreatment was very low as the MFI-UF was close to the detection limit.

At the IJssel Lake plant ultrafiltration reduced the fouling potential of the feedwater by circa 80-90% according to MFI-UF and $MFI_{0.45}$ measurements, respectively. However, the fouling potential of the final feedwater was higher than expected due to operational problems of the UF system at the time of measurement.

Minimum predicted run times based on MFI-UF measurements for a 15% flux decline to occur, assuming all particles present in the feedwater are deposited on the RO membranes, were shorter than that observed at the IJssel Lake plant. The particle deposition factor determined at both RO plants was negative, indicating negligible particle deposition and/or particles may have been sheared off the surface. The MFI-UF of tap water was observed to increase with increasing salinity in the range of the RO concentrate. An empirical correction corresponding to the ionic strength of the 85% River Rhine concentrate yielded more realistic deposition factors of close to zero. However, they remained negative for the IJssel Lake plant despite the ionic strength correction which may be due to incorrect pressure correction to account for cake compression effects in the MFI-UF measurements.

The MFI-UF determined in constant flux mode is expected to more closely simulate fouling at the surface of an RO membrane and improve the accuracy of fouling prediction by the MFI model. Preliminary experiments to determine the MFI-UF under constant flux filtration were promising, the fouling index I (and hence the MFI-UF) could be determined within a shorter time ($\approx$2 hours) compared to that normally found in the constant pressure mode. In addition, the fouling index I, determined under constant flux filtration, was demonstrated to be linear with the particulate concentration for a high and low fouling feedwater. For tap and diluted canal water, the fouling index I (and hence the MFI-UF) increased linearly with applied flux, most likely due to cake compression effects. However, further investigation of the MFI-UF in the constant flux mode at lower applied flux i.e. closer to RO operation is required.

In conclusion, the MFI-UF was found to be a promising tool for measuring the particulate fouling potential of a feedwater. It can be used alone or in combination with the $MFI_{0.45}$ or $MFI_{0.05}$ to compare the efficiency of various pretreatment processes for the removal of selected particle sizes and to determine the deposition of particles on the target membrane.

Symbols

A	membrane surface area [m²]
A_o	reference surface area of a 0.45 μm membrane filter [13.8×10⁻⁴ m²]
C_b	concentration of particles in feedwater [kg/m³]
d_p	particle diameter [m]
I	index for the propensity of particles in water to form a layer with hydraulic resistance [1/m²]
J	permeate water flux [m³/m²s]
J_o	initial permeate water flux [m³/m²s]
ΔP	applied transmembrane pressure [bar] or [N/m²]
$ΔP_c$	pressure drop over the cake [bar] or [N/m²]
$ΔP_o$	reference applied transmembrane pressure [2 bar]
R	resistance to filtration [1/m]
R_b	resistance due to blocking [1/m]
R_c	resistance of the cake [1/m]
R_m	membrane filter resistance [1/m]
r_p	diameter of particles forming the cake [m]
r	parameter with subscript applies to a RO system
t	filtration time [s]
V	filtrate volume [m³]
Y	recovery of the RO system [-]

Greek Symbols

α	specific cake resistance [m/kg]
$α_0$	constant [m/kg]
ε	cake porosity [-]
η	fluid viscosity [Ns/m²]
$η_{20°C}$	water viscosity at 20°C [Ns/m²]
$η_T$	water viscosity at temperature (T) [Ns/m²]
ψ	cake ratio factor [-]
$ρ_p$	density of particles forming the cake [kg/m³]
ω	compressibility coefficient [-]
Ω	particle deposition factor [-]

Abbreviations

CWF	clean water flux at 1 bar [l/m² h]
MFI	Modified Fouling Index using a 0.45 or 0.05 μm microfilter
MFI-UF	Modified Fouling Index using an ultrafiltration membrane
SDI	Silt Density Index

References

1.   J.G. Jacangelo, E.M. Aieta, K.E. Carns, E.W. Cummings and J. Mallevialle, Assessing hollow-fiber ultrafiltration for particulate removal, J. AWWA, (1989),81:11: 68-75.

2.   J.G. Jacangelo, J.M. Laîné, K.E. Carns, E.W. Cummings and J. Mallevialle, Low pressure membrane filtration for removing Giardia and microbial indicators, J. AWWA, (1991),83:9:97-106.

3.   S.S. Adham, J.G. Jacangelo and J.M. Laîné, Low pressure membranes: assessing integrity, J. AWWA, (1995),87:3: 62-75.

4.   P. Lipp, G. Baldauf, R. Schick, K. Elsenhans and H-H. Stabel, Integration of ultrafiltration to conventional drinking water treatment for a better particle removal-efficiency and costs. Desal., 119 (1998), pp.131-142.

5.   J.C Kruithof, J.C. Schippers, P.C. Kamp, HC Folmer and J.A.M.H. Hofman, Integrated multi-objective membrane systems for surface water treatment: pretreatment of reverse osmosis by conventional treatment and ultrafiltration, Desal., 117 (1998), 37-48.

6.   J.C. Schippers and J. Verdouw, The Modified Fouling Index, a method of determining the fouling characteristics of water, Desal., 32, (1980), 137- 148.

7.   J.C. Schippers and A. Kostense, The effect of pretreatment of River Rhine water on fouling of spiral wound reverse osmosis membranes, Proc., of the 7th International Symposium on Fresh Water from the Sea, (1980), Vol. 2 297-306.

8.   J.C. Schippers, J.H. Hanemaayer, C.A. Smolders and A. Kostense, Predicting flux decline of reverse osmosis membranes, Desal., 38, (1981), 339-348.

9.   J.C. Schippers, (1989) PhD Thesis ISBN 90-9003055-7, Vervuiling van hyperfiltratiemembranen en verstopping van infiltratieputten, Keurings instituut voor Waterleiding-artikelen Kiwa N.V, Rijswijk, The Netherlands.

10.   Ś.F.E. Boerlage, M.D. Kennedy, M.R. Dickson and J.C. Schippers, The Modified Fouling Index using ultrafiltration membranes (MFI-UF): characterisation, filtration mechanisms and membrane selection, Submitted to J. of Mem. Sci.

11.   Ś. F. E. Boerlage, M.D. Kennedy, P.A.C. Bonné, G. Galjaard and J.C. Schippers, Modified Fouling Index for the prediction of the rate of flux decline in reverse osmosis, nano and ultrafiltration systems, Proc., AWWA Membrane technology Conference, February (1997) New Orleans, USA.

12.   Ś. F. E. Boerlage, M.D. Kennedy, M. p. Aniye, E. M. Abogrean, G. Galjaard and J.C. Schippers, Monitoring particulate fouling in membrane systems, Desal., 118, (1998), 131-142.

13.   G. Green and G. Belfort, Fouling of ultrafiltration membranes: Lateral migration and the particle trajectory model, Desal.,35, (1980), 129-147.

14.   M.R. Wiesner and S. Chellam, Mass Transport considerations for pressure-driven membrane processes, J. AWWA Mem. Sci., 1 (1992), 88-95.

15.   L. Song and M. Elimelech, particle deposition onto a permeable surface in laminar flow, J. Coll. Interf. Sci., 173 (1995), 165-180.

16.   J.P. Van der Hoek, P A C. Bonné and E.A.M. Van Soest, Reverse osmosis: Finding the balance between fouling and scaling, Proceedings of the 21st IWSA World Congress September 20-26 1997, Madrid, Spain.

17.   E. Orlandini, Pesticide removal by combined ozonation and granular activated carbon filtration, Ph.D thesis, Balkema, Rotterdam, The Netherlands, (1999), ISBN 90 54104147.

18.   M.S. Chandrakanth and G.L. Amy, Effects of ozone on the colloidal stability and aggregation of particles coated with natural organic matter, Env. Sci. Tech., 30, 2 (1996), 431-443.

19.   H. Scheerman at PWN, Personal communication.

20.   I. Blume at X-Flow, Personal communication.

21.   W. Pusch Membrane Structure and its Correlation with Membrane Permeability., J. Mem. Sci., 10, (1982) 325-360.

22.   M. Cheryan, Ultra filtration Handbook, (1986), Techonomic Publishing Company Co. Inc., Lancaster, USA, 1-375.

23.   S. Chellam and M.R. Wiesner, Evaluation of crossflow filtration models based on shear induced-diffusion and particle adhesion: Complications induced by feed suspension polydispersivity, J. Mem. Sci., 138 (1998), 83-97.

24.   G. Belfort, The behaviour of suspensions and macromolecular solutions in crossflow microfiltration, J. Mem. Sci., Vol.96, (1994), 1-58.

25.   G. Foley, Modelling the Effects of Particle Polydispersity in Cross-flow Filtration. J. Mem. Sci., Vol.99, (1995), 77-88.

26   A.L. Zydney and C.K. Colton, A concentration polarisation model for the filtrate flux in cross-flow microfiltration of particulate suspensions, Chem. Eng. Commun. 47 (1985), 1-21.

27.   V. Lahoussine-Turcaud, M.R. Wiesner, and J. Bottero, Fouling in tangential flow Ultrafiltration: The effect of colloid size and coagulation pretreatment, J. Mem. Sci. **52** (1990), 173 –190.

28.   H. Mallubhotla and G. Belfort, Flux enhancement during Dean vortex microfiltration. 8. Further diagnostics, J. Mem. Sci., 125 (1997), 75-91.

29.   X. Zhu, S. Hong, A. Childress and M. Elimelech , Colloidal fouling of reverse osmosis membranes: Experimental results, Proceedings of the American Waterworks Association 1995 Membrane Technology Conference, August 13-16 Reno, U.S.A.251 -257.

30.   D. Elzo, P. Schmitz, D. Houi and S. Joscelyne, Measurement of particle/membrane interactions by a hydrodynamic method, J. Mem. Sci., 109 (1996), 43-53.

31.   P. Bacchin, P. Aimar and V. Sanchez, Influence of surface interaction on transfer during colloid ultrafiltration, J. Mem. Sci., 115 (1996), 49-63.

32.   K. Gosh and M. Schnitzer, Macromolecular structures of humic substances, Soil Sci. 129 5 (1980), 266-276.

33.   A. Braghetta, F.A. DiGiano and W. Ball, Nanofiltration of natural organic matter: pH and Ionic strength effects, J. Env. Eng. 6 (1997), 628-641.

34.   M.E. Tadros and I. Mayes, Effects of particle size, shape and surface charge on filtration, The Second World Filtration Congress, (1979), 67-72

35.   J. Kilduff and W.J.W. Weber, Transport and separation of organic macromolecules in ultrafiltration processes, Env. Sci. Technol. 26 3, (1992), 569-577.

36.   B.F. Ruth, G.H. Montillon and R.E. Montana, Studies in filtration, Ind. Eng. Chem., 25 (1933), 76.

37.   G.B. Tanny, M. Sayar, J. Rickless and M. Wilf, An automated modified fouling index device for monitoring and controlling pretreatment processes, 15th national symposium on desalination, Ashkelon, (1983).

8

Conclusions

Contents

8.1   Introduction

Pressure driven membrane filtration processes; reverse osmosis, nano, ultra and micro-filtration are undergoing a boom in growth. Due to advances in membrane manufacturing and process conditions, new applications and products are emerging and taking over market share from competing technologies. In industrial and potable water production these processes are increasingly applied for all water sources e.g. sea, ground and surface water. Similarly, in recycling and water reuse, membrane technology is playing a role, which increases in importance year by year. The choice of membrane filtration process to be applied depends on the component targeted for removal e.g. reverse osmosis for salinity, hardness and colour removal, ultrafiltration for removal of particulate matter including bacteria and viruses. Future projections are that ultrafiltration will replace conventional drinking water treatment processes on the basis of its efficiency, and economic and environmental considerations.

However, membrane scaling and fouling e.g. biofouling, organic fouling and particulate fouling are widely acknowledged as major problems associated with membrane filtration and may limit future membrane growth. This study focuses on scaling and particulate fouling as it is widely recognised that the control of these phenomena is instrumental in further membrane technology advancement and in decreasing costs associated with this process not only in drinking and industrial water production but also water recycling and reuse.

Scaling refers to the deposition of "hard scale" due to the solubility product of sparingly soluble salts being exceeded and is a problem faced mainly in reverse osmosis and nanofiltration applications. Whereas, particulate fouling is due to the deposition of suspended solids, colloids and bacteria on the membrane and is a persistent problem in all membrane filtration processes. Flux decline (reduction in water production) is an undesirable consequence of scaling and particulate fouling which results in higher energy consumption and/or reduced production capacity. Additional problems arising due to scaling and particulate fouling may be e.g. reduced membrane lifetime and increased or decreased salt retention, which translate to increased operating and cleaning costs.

Barium sulphate scaling and particulate fouling are two of the operational problems experienced at the Amsterdam Water Supply Company (AWS) RO pilot plant. Particulate fouling requires frequent cleaning to restore flux. Whereas, barium sulphate scale may be resistant to conventional RO cleaning methods if it is allowed to age into a hard deposit. Then the scale can only be dissolved by concentrated sulfuric acid and crown ethers, which would cause hydrolysis of the membrane. Therefore, prevention of barium sulphate scaling and particulate fouling is required. Scaling may be avoided by lowering the RO system recovery, although not economically desirable, or through the addition of antiscalants for sulphate and silica scales and pH adjustment for alkaline scales. However, the complete removal of particles by conventional drinking water treatment processes is not always easy or feasible, especially small colloidal particles.

The phenomena of scaling and particulate fouling are very complex and difficult to model. Presently, no quality parameter exists by which the rate of particulate fouling and scaling can be accurately predicted. Accurate prediction methods of scaling and particulate fouling could be very helpful tools for the design and operation of adequate pretreatment and for the prediction of flux decline on a routine basis. The latter aspect is important in estimating cleaning and membrane replacement frequency and ultimately, operational and pretreatment costs. Moreover, accurate scaling prediction is required to calculate the "safe" recovery and to prevent overdosing of costly antiscalants.

The goal of this research was to contribute to the understanding of the processes involved in scaling and particulate fouling in membrane filtration systems, and methods to predict and prevent these phenomena. Specific research objectives, defined in Chapter 1, were to establish (1) the solubility and kinetics of scaling in order to develop an approach for scaling prediction, using barium sulphate as a model scalant and (2) an accurate predictive test to determine the particulate fouling potential of a feedwater (further development of the Modified Fouling Index). The results of the two branches of this research *i.e.* barium sulphate scaling and particulate fouling are summarised in the following sections.

8.2   Barium Sulphate Scaling

8.2.1   Barium Sulphate Solubility Prediction

The most commonly applied method of predicting barium sulphate scaling, from the Du Pont Manual, predicts barium sulphate scaling when RO concentrate are supersaturated. Estimates using the Du Pont method at the AWS RO pilot plant indicated a scaling tendency as the RO feedwater and concentrate were predicted to be supersaturated with barium sulphate. In this chapter a solubility investigation was carried out to determine if the AWS feedwater and concentrate were really supersaturated with barium sulphate. In addition, it was aimed to develop a more accurate solubility prediction method for barium sulphate in RO concentrate which considered; ionic strength, common ion and temperature. This involved the evaluation of three different approaches, namely Du Pont, Bromley (extension of Debye Hückel) and Pitzer. Furthermore, the effect of organic matter *e.g.* humic and fluvic acids in surface water concentrate which may increase barium solubility (and hence reduce supersaturation) was examined. Barium solubility experiments in synthetic (no organic matter present) and RO surface water concentrate resulted in the following conclusions:

> The RO concentrate from the AWS RO pilot plant were confirmed to be supersaturated with barium sulphate while in general the feedwater was just undersaturated. The solubility of barium was underpredicted by the Du Pont method at it's fixed prediction temperature of 25 °C. Accurate solubility prediction was obtained at this temperature using activities calculated by the Bromley Correlation and the Pitzer Model and an experimentally determined solubility product constant (K_{sp}) for the RO concentrate. The general use of the Bromley Correlation and the Pitzer Model, with the experimentally determined K_{sp} was verified for another surface water RO concentrate from the PWN RO pilot plant. Of the two methods, the Bromley method is the easiest to apply requiring only the experimental K_{sp} and the ionic strength of the concentrate. However, when only a K_{sp} from literature is available or for more saline solutions, the Pitzer model is preferred.

Organic matter was estimated to potentially complex a negligible amount of barium by calculation and found experimentally to have no effect on barium solubility. Major influences on barium solubility were found to be ionic strength and sulphate concentration (common ion effect). The ratio of sulphate to barium (1000:1) in the AWS concentrate caused an overall decrease in barium solubility, however, the ionic strength effect results in an increase in the barium solubility of 20µg/L at 90% recovery.

To account for decreasing barium solubility (increasing supersaturation) with decreasing temperature, a relationship was derived from data available in literature for barium solubility in pure water solutions at different temperatures. This relationship was used to to correct the experimentally determined K_{sp} for temperature. This relationship was verified using AWS RO and synthetic concentrate for the 5 - 25°C temperature range. Use of the temperature corrected K_{sp} and the Pitzer model allowed accurate prediction at temperatures relevant to the drinking water industry for surface water treatment in the Netherlands.

According to the Bromley Correlation, the Pitzer Model and the experimentally determined solubilities, the concentrate of the RO system at AWS are significantly supersaturated and therefore, thermodynamically, scaling could occur. However, in practice no scaling occurred at 80% recovery and the results of this chapter confirmed that a metastable zone existed in barium sulfate supersaturated solutions for AWS concentrate. This zone was found to be quite wide, *i.e.* at least 27 times the solubility limit at 5°C.

8.2.2   Stable Barium Sulphate Supersaturation

The stable barium sulphate supersaturation phenomenom confirmed in Chapter 1 for the RO concentrate at AWS, may be explained by slow precipitation kinetics i.e. nucleation and growth and/or inhibition of kinetics by organic matter present in the concentrate. The objective of this chapter was to identify the cause of stable supersaturation in the AWS RO concentrate from a study of barium sulphate precipitation kinetics. In examining nucleation i.e. the birth of a crystal, the concept of induction time was used to estimate nucleation time. It was assumed that the time required for nucleation was much greater than the time required for growth of crystal nuclei to a measurable size. The growth rate of barium sulphate crystals in the RO concentrate was determined from seeded growth experiments. Experiments to determine the induction time and growth rate in AWS RO concentrate and synthetic concentrate containing (i) no organic matter and (ii) commercial humic acid (chosen as a model organic compound) resulted in the following conclusions:

Supersaturation (expressed as supersaturation ratio, S_r) was found to play a major role in determining the barium sulphate induction time and hence the stability of the RO concentrate. Measured induction time decreased more than 36 times with a recovery increase from 80% to 90%, corresponding to a supersaturation of 3.1 and 4.9, respectively. Using data published in literature for barium sulphate induction time, temperature was also found to have a significant effect on induction time which was most pronounced at lower supersaturation. For example, for the temperature range experienced at the AWS RO pilot plant of 5 - 25°C, a temperature increase of 20°C was expected to shorten induction time by a factor of 3.6 at a supersaturation ratio of 5.

Organic matter present in the RO concentrate did not prolong nucleation time. Whereas, addition of commercial humic acid (20 mg/L) to the 90% synthetic concentrate extended the induction time by more than 50 times. However, from the results of the growth experiments it was concluded that this was most likely due to inhibition of growth to a detectable size in the induction time measurement.

The barium sulphate growth rate in the RO concentrate generally increased with increasing recovery due to the higher supersaturation i.e. driving force. However, the measured growth rates were 60-120 times lower than those reported in literature for pure water solutions of a similar supersaturation. This was due in part to the inhibitory effect of organic matter present in the RO concentrate, as the initial growth rate in the 90% RO concentrate was a factor 2.5 lower than in the synthetic concentrate. In addition, most likely the 1000:1 sulphate to barium ion stoichiometry and the higher ionic strength of the RO concentrate also contributed to the lower growth rates measured for the RO concentrate. Despite the inhibitory effect of organic matter, growth was not totally arrested and overgrowth of the adsorbed organic matter allowed barium desupersaturation. Commercial humic acid was more effective with the growth rate six times lower in comparison to the synthetic concentrate. Moreover, total growth inhibition was observed in 80% synthetic concentrate after 25 hours. However, at higher supersaturation (90% synthetic concentrate), overgrowth of humic acid also occurred, leading to barium desupersaturation.

In conclusion, inhibition of nucleation by organic matter present in the RO concentrate was not found to limit barium sulphate precipitation. However, the induction time observed at 80% recovery was very long (>200hours) and this most likely prevented scaling in the RO system. In contrast, the growth rate measured for 80% RO concentrate was significant. If the growth rate was limiting precipitation then scaling should have occurred at 80% recovery in the pilot plant, especially over the prolonged period of operation at this recovery. However, this was not the case and thus, growth was not the factor governing scaling. Instead results indicated stable supersaturation in the RO concentrate was due to long induction times (slow nucleation kinetics).

8.2.3   Barium Sulphate Scaling Prediction

The final objective of the barium sulphate scaling research was to develop a method which more realistically predicts under which supersaturation and temperature conditions, barium sulphate scaling occurs in the RO system. The stability (i.e. no scaling) of RO concentrate supersaturated with barium sulphate, was attributed to long induction times (slow nucleation kinetics) in the previous chapter. Therefore, in developing this scaling prediction method, it was assumed that a threshold induction time could be defined which should not be exceeded in order to prevent scaling in the RO system. Induction time is the net effect of RO feedwater temperature and supersaturation. As was shown from the previous chapter, induction time is shorter with increasing supersaturation and temperature. However, temperature has a double effect whereby a temperature increase also decreases the supersaturation and hence extends the induction time. Therefore, in developing this scaling prediction method the following steps were carried out; a comparison of methods to calculate the supersaturation ratio for the RO concentrate i.e. the existing Du Pont method and the Pitzer model (refer Chapter 1). Subsequently, induction times of the RO concentrate were calculated for the supersaturation and temperature conditions occurring during operation from measured induction times of the AWS RO concentrate at 25 °C (refer Chapter 2). Finally, an analysis of the calculated induction times for the various operating modes during which no scaling and scaling was found in the RO system was performed. The findings of this chapter were as follows:

The temperature limitation of the Du Pont method to 25 °C coupled with solubility being underestimated at 25 °C led to an error in quantifying supersaturation in the RO concentrate. Supersaturation in the RO concentrate was underestimated by 40% for the lowest RO feedwater temperature of 3.3 °C and overestimated by +15% at 25 °C. Therefore, the more accurate activity method (the Pitzer model and the experimentally determined solubility product constant, K_{sp}) was employed to quantify supersaturation which corrects for variations in the RO feedwater temperature. Using this method it was shown that the supersaturation could vary dramatically for a given recovery (and hence the scaling potential) due to variable feedwater concentration and temperature. For example the supersaturation ratio (S_r) showed a three fold difference during operation at 90% recovery from a S_r 5.4 to 17.5 .

The median induction times calculated for the 90% RO concentrate during operation with addition of sulfuric acid and either antiscalant A or B and (modes 1 and 2, respectively) were very short at only 18 minutes. Scaling episodes during operation with antiscalant A coincided with shorter induction times of 2-15 minutes. The failure of this antiscalant suggested that induction times in this range most likely will result in scaling. Therefore, operating within this range requires the addition of a more effective antiscalant. Although, antiscalant B was effective it caused serious biofouling.

The third operation mode operated at AWS; 80% recovery with hydrochloric acid addition and no antiscalant reduced the median supersaturation by more than half. This extended the median induction time to >82hours, more than 270 times longer than during operation at the higher recovery of 90% and with sulfuric acid addition (modes 1 and 2). No scaling was evident despite the high supersaturation during nineteen months of operation in this mode.

Safe and unsafe induction time limits, were determined by correlating induction time to periods when scaling did and did not occur in the RO system during operation mode 4. In this mode the recovery was increased step wise from 86, 88 to 90% recovery. Induction times shorter than 5 hours most likely resulted in scaling while induction times greater than 10 hours were expected to be safe. Safe long term operation at 80 and 85% recovery was attributed to operation with 90% and 66% of the induction times above the safe induction time limit, respectively.

Finally, safe and risky supersaturation limits for the AWS RO pilot plant were derived corresponding to induction times of 10 and 5 hours, respectively. In the temperature range of 5-25°C, these supersaturation limits range from 5.4-4.6 (safe) to 6.0-5.0 (risky). Use of these limits gives a more dynamic approach to avoiding scaling while optimising the RO system recovery. This involves monitoring the supersaturation and temperature over time which allows the scaling potential to be assessed and recovery adjustment to maintain operation within these limits. However, the validity of the suggested induction times and corresponding supersaturation limits needs to be confirmed with pilot studies using different reverse osmosis elements and with feedwater of different quality at various recoveries before they can be considered for general use.

8.3   Particulate Fouling

8.3.1   Development of the MFI-UF Index

An investigation of the application of hollow fibre ultrafiltration (UF) membranes in the existing Modified Fouling Index ($MFI_{0.45}$ or $MFI_{0.05}$) test was the first objective of the particulate fouling research. The development of the new MFI-UF index was aimed at incorporating the fouling potential of smaller colloidal particles not measured by the existing $MFI_{0.45}$ or $MFI_{0.05}$ tests. In addition to developing the MFI-UF, one of the research objectives of this chapter was to propose a suitable reference membrane for the MFI-UF test. Important UF membrane properties in choosing a reference membrane included; membrane material, pore size expressed as molecular-weight-cut-off (MWCO) and surface morphology characteristics e.g. surface porosity as these properties may affect the occurrence of cake filtration (the basis of the MFI test) and the MFI-UF value obtained. A final consideration is that small UF membrane elements are expensive. Therefore, for the test to be viable in practice, a simple method is required to clean the UF membranes to allow membrane reuse These aspects were examined in tap water experiments using ultrafiltration membranes of a broad MWCO range (1 to 100 kDa) manufactured from two different membrane materials (polysulphone and polyacrylonitrile). Field emission scanning electron microscopy (FESEM) of the membrane surfaces was employed to determine surface morphology characteristics. The findings of this chapter were as follows:

> For polyacrylonitrile membranes the (real) MFI-UF* value (indicated as MFI-UF*) was obtained within 20-50 hours of filtration when the MFI-UF value stabilised over time. In contrast, the MFI-UF value for polysulphone membranes continuously decreased over time. Furthermore, polyacrylonitrile (PAN) membranes were easier to clean, as one cleaning with sodium hypochlorite restored the clean water flux to ≈100%. Whereas, polysulphone (PS) membranes needed repeated chemical cleaning to restore the clean water flux.

The MFI is dependent on particle size through the Carmen-Kozeny equation for specific cake resistance. The application of UF membranes in the MFI test resulted in the retention of smaller particles as expected. This was evident by the significantly higher MFI-UF* values (2000-13 300 s/l^2) in comparison to the $MFI_{0.45}$ (1-5 s/l^2) commonly measured for tap water. However, the MFI-UF* obtained appeared rather independent of MWCO as the MFI-UF* varied to only a small extent, from 2000 to 4500 s/L^2 for PAN and PS membranes ranging in MWCO from 3 to 100 kDa. Only the 1, 2, 5 kDa polysulphone membranes from manufacturer A gave markedly higher MFI-UF* values of 8400 - 13 300 s/L^2.

The polyacrylonitrile membrane surfaces were shown to be homogeneously permeable from FESEM micrographs. Although, the porosity and pore size ranking determined from FESEM of 13 kDa < 6 kDa < 50 kDa did not correspond with the increasing MFI-UF* value in the order PAN 6 kDa < 13 kDa ≈ 50 kDa. The lower MFI-UF* value obtained with the 6 kDa membrane was attributed to its low flux and as a result of lower cake compression.

FESEM micrographs of the 1, 2, 5 kDa polysulphone membranes from manufacturer A showed them to be heterogeneously porous due to narrow striations running across the membrane where pores were located. In between these striations, very wide and very narrow bands were observed where no pores were present. Consequently, the much higher MFI-UF* values found for these membranes were attributed to their lower active filtration area and not due to the retention of smaller particles. In contrast, no impermeable striations were found on the surfaces of the other polysulphone membranes.

Cake filtration was the dominant filtration mechanism for the polyacrylonitrile membranes, demonstrated by linearity in the t/V versus V plot. Whereas, for the polysulphone membranes, especially the 1, 2, 5 kDa membranes, cake filtration was difficult to demonstrate. The changing gradient in the t/V versus V plots of these membranes was attributed to their low surface porosity and/or the irregular distribution of pores over the membrane surface. These surface characteristics were thought to cause a significant lateral flow component in the streamline flow and a steady increase in effective filtration area over time. Thus, the steadily decreasing MFI-UF value observed over time. Once these effects can be neglected, most likely the cake acts as a second membrane, determining the size of particles retained and hence the resultant MFI-UF* value obtained. This explains why the MFI-UF* appeared rather independent of MWCO for the 3 - 100 kDa membranes

The PAN 13 kDa membrane was proposed as a suitable reference membrane for application in the MFI-UF test. Cake filtration was proven to occur which gave a stable MFI-UF* value. Furthermore, the pores of the PAN 13 kDa membrane are circa 1000 times smaller than the pores of the existing $MFI_{0.45}$ test membrane and thus will include much smaller particles in the measurement.

8.3.2   The MFI-UF as a Water Quality Test and Monitor

In Chapter 6 various aspects of the new MFI-UF test, employing the reference polyacrylonitrile (PAN) 13 kDa membrane proposed in the previous chapter, were investigated to establish its general use for characterising the fouling potential of feedwater. The objectives of this research included; firstly, verifying the occurrence of cake filtration and a linear relationship between the MFI-UF index to particle concentration for water of varying quality. Secondly, demonstrating the reproducibility of the MFI-UF value when using different PAN 13 kDa membranes and with repeated use of one PAN 13 kDa membrane with membrane cleaning between measurements. Thirdly, determining the effect of process conditions in the MFI-UF test i.e. the pressure and temperature dependency of the specific cake resistance measured in the MFI-UF index and that of MFI-UF reference membrane. Finally, the application of the MFI-UF test as a monitor to detect feedwater changes over time was examined. Experiments using feedwater of varying fouling potential resulted in the following conclusions:

> Cake filtration was shown to occur for high and low fouling feedwater by the stability of the MFI-UF value over time. The MFI-UF value was found to be stable with prolonged filtration, up to circa 60 hours for WRK-I water (a conventionally pretreated surface water). Whereas, cake filtration in the existing $MFI_{0.45}$ test occurred for a considerably shorter period of time ≈ 10 minutes for WRK-III water (also a conventionally pretreated surface water).

> The MFI-UF index was demonstrated to be linear with particulate concentration for all the feedwater (high and low fouling feedwater) tested and at a range of applied test pressures which further supports that cake filtration is occurring in the MFI-UF test.

> The MFI-UF value obtained for the same feedwater was reproducible for 83% of the membranes tested from three different batches of manufacture. In addition, a reproducible MFI-UF value was found in all five tests using the same membrane, applying chemical cleaning in between each measurement to restore the clean water flux to 100%.

> The PAN 13 kDa membrane resistance was found to be temperature independent. In addition, the specific cake resistance measured in the MFI-UF index after correction for feedwater viscosity was found to be temperature independent for the 15-31 °C temperature range examined. Therefore, temperature correction in the MFI-UF test requires only a feedwater viscosity correction from the ambient test temperature to the standard reference temperature of 20 °C.

The observed compaction of the PAN 13 kDa membrane with increasing applied pressure (0 to 2 bar range) was only minor (8%) and therefore, does not need to corrected for. In contrast, the specific cake resistance measured in the MFI-UF index was found to increase significantly with an increase in applied pressure due to cake compression effects. The compressibility coefficient determined for a range of feedwater was *circa* 0.51-0.85. The average of these individual feedwater compressibility coefficients (0.75) was used as a global compressibility coefficient to account for cake compression in the pressure correction factor. This resulted in 76% of the MFI-UF measurements within ±20% of the true MFI-UF value experimentally determined at the reference pressure of 2 bar. While the use of the individual feedwater compressibility coefficients, was marginally better with 82% of the MFI-UF measurements within ±20% to the true MFI-UF value.

The MFI-UF test could not be applied to quantify the fouling potential of a highly variable feedwater over time i.e. operate as a monitor, as the resultant MFI-UF value may be due to the combination of depth filtration and compression effects in addition to cake filtration. Moreover, the delayed response in the MFI-UF index to a change in feedwater, may be due to the history effect in the calculation of the MFI-UF via the t/V vs V plot. Consequently, application of the MFI-UF as a monitor requires further investigation.

8.3.3   Applications of the MFI-UF

The last objective in the particulate fouling research was to verify that the MFI-UF test could be applied to measure and to predict the particulate fouling potential of reverse osmosis (RO) feedwater. In addition, the MFI-UF test could also be applied to assess the efficiency of pretreatment processes for the removal of smaller particles (in the size range of " $\geq 13\,000$ kDa") than the existing fouling indices namely the $MFI_{0.45}$ and SDI. Furthermore, it was aimed to extend the existing MFI model for fouling prediction to the MFI-UF index. The MFI model assumes that the fouling process in an RO system follows conventional cake filtration behaviour. Parameters to improve fouling prediction in the model investigated in this chapter included the following: determining the particle deposition factor at the AWS (River Rhine) and PWN (IJssel Lake) RO pilot plants, the effect of increasing salinity on cake resistance in tap water and the development of the MFI-UF test in the constant flux mode. Experiments using tap water, RO feedwater after extended pretreatment processes and RO concentrate at two RO pilot plants yielded the following conclusions:

The MFI-UF of the influent feedwater to the RO Pilot Plants was circa 700-1900 times higher than that of the respective $MFI_{0.45}$ due to the retention of smaller particles " $\geq 13\ 000$ kDa"in the MFI-UF test. Slow sand filtration was the most successful pretreatment process at the River Rhine Plant with consistently 90-$\approx$100% removal of particles in the size range measured by the MFI-UF and $MFI_{0.45}$ tests. Whereas, the SDI gave variable results for this step at 40->90%. The fouling potential of the RO feedwater at this plant after extended pretreatment was very low as the MFI-UF was close to the detection limit (100 s/l^2). At the IJssel Lake plant, ultrafiltration reduced the fouling potential of the feedwater by circa 80-90% according to MFI-UF and $MFI_{0.45}$ measurements, respectively. However, the fouling potential of the final feedwater was higher than expected most likely due to operational problems of the UF pilot system at the time of measurement.

Minimum predicted run times at the IJssel Lake RO plant based on MFI-UF measurements for a 15% flux decline to occur, assuming all particles present in the feedwater were deposited on the RO membranes, were shorter than that observed. The particle deposition factor determined at both RO plants was negative, suggesting negligible particle deposition and/ or particles may have been sheared off the surface. The MFI-UF of tap water was observed to increase with increasing salinity in the range of the RO concentrate. An empirical correction corresponding to the ionic strength of the 85% River Rhine concentrate yielded more realistic deposition factors of close to zero. However, they remained negative for the IJssel Lake plant despite an ionic strength correction which may be due to incorrect pressure correction to account for cake compression effects in the MFI-UF measurements.

The MFI-UF determined in constant flux mode was expected to more closely simulate fouling at the surface of an RO membrane and improve the accuracy of fouling prediction by the MFI model. Preliminary experiments to determine the MFI-UF under constant flux filtration were promising, the MFI-UF could be determined within a shorter time ($\approx$2 hours) compared to that normally found in the constant pressure mode. The MFI-UF index was demonstrated to be linear with particulate concentration also in this mode for a low fouling feedwater (tap water) and a high fouling feedwater (diluted canal water). For these feedwater, the fouling index I (and hence the MFI-UF) increased linearly with applied flux, in the range of 70 to 110 l/m^{2}hr, most likely due to cake compression effects. However, further investigation of the MFI-UF in the constant flux mode at lower applied flux i.e. closer to RO operation is required.

In conclusion, the MFI-UF (measured at constant pressure or constant flux) was found to be a promising tool for measuring the particulate fouling potential of a feedwater. It can be used alone or in combination with the $MFI_{0.45}$ or $MFI_{0.05}$ to compare the efficiency of various pretreatment processes for the removal of selected particle sizes and to determine the deposition of particles in the target membrane systems.

Samenvatting

In de afgelopen 10 jaar vertoont de toepassing van membraanfiltratie; omgekeerde osmose, nano-, ultra-, en microfiltratie een continue groei. Deze groei wordt gestimuleerd door inzet van ontzouting in gebieden met toenemend tekort aan water en ook de stringentere wetgeving die een verdergaande behandeling van drinkwater vergt, zoals bijvoorbeeld de verwijdering van Giardia en Cryptosporidum cysten volgens de Surface Water Treatment Rule (V.S.). Innovaties in de produktie van membranen en procescondities hebben geresulteerd in een aanzienlijke daling van de kosten van membraanfiltratie processen. Als gevolg daarvan kan membraanfiltratie qua kosten concurreren met conventionele methoden en is dan een alternatief bij de bereiding van drink en industrieel water en bij hergebruik van water. De inzet van membraanfiltratie om onze waterkwaliteitsproblemen op te lossen staat nog in de kinderschoenen, te meer daar nieuwe producten en toepassing en in ontwikkeling zijn. Echter precipitatie van anorganische verbindingen (scaling), groei van bacteriën, adsorptie van organische stoffen en afzetting van zwevende en collodale deeltjes (in dit proefschrift worden scaling en vervuiling door deeltjes behandeld) doen zich gelden als serieuze belemmeringen bij toekomstige groei van de membraantechnologie. Scaling, dat vooral optreedt bij omgekeerde osmose en bij nanofiltratie, is het gevolg van de vorming van harde afzettingen op het membraan en is te wijten aan de beperkte oplosbaarheid van verschillende verbindingen, zoals bijvoorbeeld $BaSO_4$. Vervuiling door deeltjes is een bijzonder hardnekkig probleem bij alle membraanfiltratieprocessen en heeft betrekking op de afzetting van zwevende stoffen, colloïdaal materiaal en micro-organismen op het membraan. De problemen die voortkomen uit scaling en uit vervuiling door deeltjes leiden tot een afname in de filtratiesnelheid (flux) of tot een hogere benodigde druk om de flux te handhaven, hetgeen leidt tot een toename van de operationele kosten. Het reinigen van het membraan met als doel de verwijdering van afzettingen van anorganische verbindingen, deeltjes en andere stoffen resulteert in een verlenging van stilstand waarin niet wordt geproduceerd, toename van energie en het chemicaliëngebruik en leidt ook tot het ontstaan van afvalwater en brengt daardoor hogere kosten met zich mee. Wanneer reiniging van de membranen niet effectief is, dan zullen deze moeten worden vervangen om de productiecapaciteit te waarborgen.

Het is algemeen bekend dat het beheersen van scaling en vervuiling door deeltjes essenteel is voor de verdere ontwikkeling van de membraantechnologie en verlaging van de kosten van menbraantechnologie. Dit kan alleen worden bereikt wanneer op betrouwbare wijze wordt vastgesteld of scaling en afzetting van deeltjes te verwachten is. Om die reden is het gebruikelijk om eerst proefinstallatie onderzoek te verrichten alvorens een praktijkinstallatie te ontwerpen. Hoewel dit in het algemeen een goed beeld geeft van de situatie, kost het veel tijd en dus veel geld. Het doel van dit onderzoek was dan ook om methoden te ontwikkelen om scaling (gebruikmakend van bariumsulfaat als model scalant) en vervuiling door deeltjes in membraanfiltratie systemen te voorspelen. Deze methoden kunnen dan als gereedschap worden toegepast om vast te stellen en te monitoren of de maatregelen en verbeteringen hiervan om scaling en afzetting te voorkomen, effectief zijn, en wel zonder gebruik te maken van duur onderzoek met proefinstallaties met het doel tijd en kosten te besparen.

Hoofdstuk 1 van de proefschrift geeft een overzicht van membraanfiltratie bij de drink- en industriewaterbereiding en beschrijft de meest voorkomende scalants en vervuiling, als ook de bestaande methoden om deze verschijnselen te voorspellen en te beheersen. Ook worden tekortkomingen van bestaande methoden bij het voorspellen van scaling en vervuiling aangegeven.

Bij één omgekeerde osmose proefinstallatie in Nederland, waarmee water afkomstig uit de Rijn is behandeld, bleek ondanks de toevoeging van antiscalant, zich toch bariumsulfaat af te zetten. Terwijl onder andere procescondities, zonder toevoeging van antiscalant, geen scaling optrad, hoewel de voorspelde oververzadiging en dus de neiging tot afzetting hoog was. Een vergelijkbare situatie doet zich voor bij de meest algemeen toegepaste methode om de vervuilingspotentie door deeltjes te meten, te weten de de Silt Density Index (SDI) en de Modified Fouling Index ($MFI_{0.45}$). Beide methoden simuleren de membraanvervuiling door het water te filteren door een filter met porieen van $0.45\mu m$ bij constante druk en laten daardoor kleinere deeltjes buiten beschouwing. Hierbij komt dat de SDI niet is gebaseerd op enig filtratie-mechanisme en dus ook geen lineair verband vertoont met de concentratie van colloidale/zwevende deeltjes. Hierdoor is het niet mogelijk de SDI te gebruiken als basis voor een model dat de afname van de flux, in bijvoorbeeld een omgekeerde osmose installatie voorspelt. In tegenstelling tot de SDI, is de $MFI_{0.45}$ gebaseerd op koekfiltratie en rechtevenredig met het gehalte aan deeltjes en kan als zodanig wel dienen als basis voor een model. De correlatie met de praktijk is echter onvoldoende omdat de $MFI_{0.45}$ ongevoelig is voor de kleinere deeltjes die hoogst waarschijnlijk verantwoordelijk zijn voor de vervuiling. Het doel van het onderzoek omvat twee richtingen te weten: 1) de oplosbaarheid en kinetiek van scaling en de ontwikkeling van een aanpak waarmee het optreden van scaling kan worden voorspeld (waarbij bariumsulfaat als model scalant is gebruikt) en 2) een test waarmee de vervuilingspotentie van water betrouwbaar kan worden voorspeld (verder ontwikkelen van de MFI door toepassing van ultrafiltratie membranen met kleinere porieën). Vervolgens zijn deze methoden toegepast, om vast te stellen of getroffen maatregelen om scaling en vervuling te vookomen effectief waren.

In hoofdstuk 2 wordt de betrouwbaarheid van de meest toegepaste methode om bariumsulfaat scaling te voorspellen, te weten, het Du Pont Manual, onderzocht. Deze methode voorspelde dat de bariumoplosbaarheid van het concentraat in de omgekeerde osmose proefinstallatie van de Gemeentewaterleidingen Amsterdam (GWA) was overschreden met een factor 14 bij een 80% opbrengst berekend bij 25°C. Toch trad er geen scaling op in deze proefinstallatie gedurende meer dan een jaar bij een opbrengst van 80%. Mogelijke verklaringen hiervoor zijn: een onnauwkeurige voorspelling van de oplosbaarheid bijvoorbeeld omdat het concentraat van de omgekeerde osmose installatie niet echt oververzadigd was en/of organische stoffen vormden complexen met barium. Dit laatste is onderzocht door de oplosbaarheid van barium in het concentraat van de omgekeerde osmose installatie te bepalen en in een kunstmatig bereid concentraat (zonder organische stoffen). Hieruit bleek dat barium inderdaad oververzadigd was en dat organische stoffen geen effect hadden op de oplosbaarheid. De door DuPont gehanteerde methode gaf een 30% lagere oplosbaarheid onder dezelfde condities. Tenslotte is een nauwkeurige methode ontwikkeld en geverifieerd voor de berekening van de oplosbaarheid van bariumsulfaat (en om de oververzadiging te kwantificeren) in het concentraat van omgekeerde osmose installaties in een temperatuur gebied van 5-25°C. Deze methode maakt gebruik van Pitzer activiteitscoëfficiënten in een experimenteel vastgestelde oplosbaarheidsproduct (K_{sp}) in het concentraat.

In hoofdstuk 3 wordt de oorzaak van een permante oververzadiging in het concentraat van de proefinstallatie van GWA zonder dat bariumsulfaat neerslag. Tevens is de trage kinetiek en/of remming van de precipitatie door organische stof onderzocht. De kinetiek van de precipitatie van bariumsulfaat; nucleatie van kristallen, gemeten als inductietijd en groei, zijn onderzocht in ladingsgewijze experimenten in concentraat van omgekeerde osmose installaties en in kunstmatig bereid concentraat. Dit laatste zonder organische stoffen en met organische stoffen in de vorm van commercieel verkrijgbaar humuszuur. Oververzadiging blijkt in hoge mate de inductietijd te bepalen. De inductietijd werd meer dan 36 maal verkort wanneer de opbrengst werd verhoogd van 80% naar 90%, hetgeen overeenkomt met een oververzadiging van respectievelijk 3.1 en 4.9. Organische stoffen in het concentraat van de omgekeerde osmose installatie bij 90% opbrengst gaf geen wezenlijk verlenging van de inductietijd (5,5 uur). Terwijl toevoeging van commercieel verkrijgbaar humuszuur de inductie tijd verlengde tot 200 uur. Dit effect kan waarschijnlijk worden toegeschreven aan vertraging van de kristalgroei, daar groeisnelheden bepaald met entkristallen in kunstmatig bereid concentraat waaraan verkrijgbaar humuszuur was toegevoegd, met een factor 6 werden verlaagd. Groeisnelheden, in concentraat van de omgekeerde osmose installatie, werden door organische stoffen slechts 2,5 maal vertraagd. Echter de gemeten groeisnelheden in concentraten bij 80 en 90% opbrengsten bleken toch nog opmerkelijk hoog te zijn en vrijwel zeker niet de precipitatie van barium te kunnen verhinderen. De resultaten wijzen in de richting dat de nucluatiesnelheid, gemeten als inductietijd, het optreden van scaling bepaalt.

In hoofdstuk 4 wordt een methode ontwikkeld om het optreden van scaling door bariumsulfaat te kunnen voorspellen, gebaseerd op de veronderstelling dat een drempelwaarde voor de inductietijd kan worden gedefinieerd waarboven geen scaling optreedt. Inductietijden zijn berekend voor oververzadigingswaarden (bepaald met het Pitzermodel) en temperatuurgegevens van de proefinstallatie van GWA. Hierbij tevens gebruikmakend van een verband tussen inductie tijden en temperatuur. Veilige (> 10 uur) en onveilige (< 5 uur) grenswaarden voor inductietijden zijn afgeleid van perioden waarin wel en geen scaling optrad bij opbrengsten van 80 tot 90%. Gebaseerd op deze inductietijden zijn grenswaarden (veilige en onveilige) afgeleid voor een temperatuurgebied van 5-25°C. Gebruik van deze grenswaarden opent de mogelijkheid tot een meer flexibele bedrijfsvoering bij de optimalisatie van de opbrengst in installaties, terwijl scaling wordt vermeden. De algemene geldigheid van deze grenswaarden dient verder te worden geverifieerd met voedingwater van verschillende kwaliteit en met gebruikmaking van verschillende typen membraanelementen.

In hoofdstuk 5 wordt de ontwikkeling van de Modified Fouling Index beschreven, waarbij ultrafiltratiemembranen worden gebruikt (MFI-UF index). Bij deze index wordt ook de bijdrage van colloïdale deeltjes aan de vervuilingspotentie van het water gemeten, hetgeen niet of slechts, ten dele gebeurt bij de $MFI_{0.45}$ en $MFI_{0.05}$. Om tot een voorstel te komen voor een geschikt membraan dat als referentie kan dienen, zijn polysulfon en polyacrilonitril membranen getest, met verschillende poriëngrootte (molecular-weight-cut-off van 1-100 kDa MWCO) op kraanwater. De gemeten MFI-UF waarden (2000-13300 s/L^2) blijken aanzienlijk hoger dan die voor de $MFI_{0.45}$ (1-5 s/L^2). Dit wijst duidelijk op de aanwezigheid van kleinere deeltjes daar de MFI afhankelijk is van de deeltjesgrootte zoals wordt beschreven in de Carmen-Kozeny vergelijking. Echter de MFI-UF blijkt onafhankelijk te zijn van de MWCO, in het gebeid van 3-100 kDa; dit waarschijnlijk omdat de gevormde koek zich als een tweede membraan met fijnere poriën gaat gedragen.

Het polyacrilonitril membraan met 13 kDa MWCO wordt voorgesteld als het meest geschikte referentiemembraan voor de MFI-UF test. Dit van wege het feit dats koekfiltratie, de basis van de MFI-test, het dominante filtratiemechanisme is, hetgeen blijk uit het lineaire verband tussen t/V versus V. Dit geeft als resultaat een stabiele MFI-UF waarde gemeten in de tijd. Tevens laten FESEM opnamen (field emission scanning electron microscopy) van het membraan zien dat poriën circa 1000 maal kleiner zijn dan de poriën in de membranen die gebruikt worden voor de $MFI_{0.45}$ test.

In hoofdstuk 6 worden verschillende aspecten van de nieuw ontwikkelde MFI-UF test onderzocht; dit om de algemene toepasbaarheid vast te stellen, zoals: het bewijs dat koekfiltratie inderdaad optreedt, het lineaire verband tussen de index en de concentratie van de zwevende en colloidale deeltjes zowel in water met een geringe als een hoge vervuilingspotentie van het water, de reproduceerbaarheid van de test, herleiding van de meetwaarden naar standaard condities (2 bar en 20°C) en de toepassing als monitor om continu de vervuilingspotentie van water kunnen meten. Koekfiltratie is aangetoond op te treden bij testen met water met een geringe en hoge vervuilingspotentie. De test bleek reproduceerbaar voor 83% van de beproefde membranen die afkomstig waren van drie verschillende batches en in vijf testen waarbij hetzelfde membraan na reiniging weer is gebruikt. Herleiding van de meetwaarden verkregen bij andere temperaturen dan de referentie waarde (20°C) vergt slechts een correctie die is gebaseerd op de viscositeit. Echter de gevormde koek op de membraan blijkt bij alle water typen, te worden samengedrukt, afhankelijk van de gebruikte druk. Daarom zijn de compressibiliteitscoëfficiënten bepaald voor verschillende water types en is een algemene coëfficiënt berekend om de MFI-UF te kunnen herleiden tot de referentie druk.

De huidige MFI-UF test is niet geschikt om de vervuilingspotentie continu te meten. Dit, waarschijnlijk door het optreden van een combinatie van het optreden van koekfiltratie en dieptefiltratie en/of samendrukking van de koek. Bovendien is er een trage response van de MFI-UF waarden bij verandering van de waterkwaliteit, die mogelijk wordt veroorzaakt door de invloed van voorgaande meetwaarden bij de berekening van de MFI-UF uit de relatie t/V versus V. Nauwkeuriger meting van de tijd en het gefilteerde volume kunnen dit probleem mogelijk oplossen. Echter, de resultaten in dit hoofdstuk laten zien dat deze test gebruikt kan worden om een momentopname te maken van der vervuilingspotentie van water.

In hoofdstuk 7 wordt de MFI-UF toegepast om de vervuilingspotentie van voedingswater voor omgekeerde osmose installaties te meten en te voorspellen. MFI-UF metingen zijn uitgevoerd bij constante druk bij de proefinstallaties die worden gevoed door water afkomstig van het IJsselmeer en de Rijn. Dit zowel voor het voedingswater als na verschillende voorbehandelingen, zoals coagulatie, sedimentatie, snelfiltratie, ultrafiltratie enz. Het effect van de voorbehandeling is beoordeeld op basis van de MFI-UF en vergeleken met de $MFI_{0.45}$ waarden. De MFI-UF van het voedingswater bleek 700-1900 maal hoger te zijn dan de corresponderende $MFI_{0.45}$ waarden. De effectiviteit van de voorbehandeling gebaseerd op de MFI-UF was voor beide installaties 80%. Voor de grotere deeltjes uitgedrukt als $MFI_{0.45}$ was de effectiviteit zelfs 90-100%. Voorspelde minimum looptijden, van omgekeerde osmose installaties gevoed met dit type water, alvorens een flux afname van 15% wordt bereikt, blijken duidelijk korter te zijn dan gerealizeerd in de proefinstallaties. Hoogstwaarschijnlijk werd dit verschil veroorzaakt door bijna verwaarloosbare aftzetting van deeltjes op de membranen en/of verwijdering van de koek die gevormd onder de langstroom condities in de membraanelementen.

Bovendien is aangetoond dat de weerstand van de koek toeneemt bij toenemende ionsterkte in kraanwater. Een correctie van de MFI-UF voor de ionsterkte in membraan concentraat is dan ook nodig. Tenslotte wordt voorgesteld de MFI-UF bij constante flux te meten omdat hierbij de condities in omgekeerde osmose installaties dichter worden benaderd. Orienterende experimenten waren zeer veel veelbelovend, de MFI-UF kan n.l. worden vastgesteld binnen 2 uur. Dit zowel voor water met een geringe als een hoge vervuilingspotentie. En er blijkt een lineair verband te bestaan tussen de MFI-UF en de concentratie van de zwevende/colloidale deeltjes. Concluderend kan worden gesteld dat de MFI-UF (gemeten bij constante druk of flux) een veelbelovend hulpmiddel is om de vervuilingspotentie door deeltjes van water te kunnen meten. Dit kan alleen of in combinatie met de $MFI_{0.45}$ worden uitgevoerd, bijvoorbeeld om het effect van de verschillende stappen in de voorbehandeling van voedingwater van membraanfiltratie-installaties te meten en de aftzetting van deeltjes in een membraansysteem te kwantificeren. Hierbij kan onderscheid worden gemaakt tussen deeltjes van verschillende grootte.

In hoofdstuk 8 worden de belangrijkste conclusies van het onderzoek samengevat.

List of Publications

1.   Ś.F.E. Boerlage, M.D. Kennedy, G.J. Witkamp, J.P. Van der Hoek and J.C. Schippers. Prediction of BaSO$_4$ solubility in reverse osmosis systems, Proceedings of the American Waterworks Association (1997) Membrane Technology Conference, February 23-26, New Orleans, U.S.A. 745-759.

2.   Ś.F.E. Boerlage, M.D. Kennedy, P.A.C. Bonne, G. Galjaardand J.C. Schippers, Modified Fouling Index for the prediction of the rate of flux decline in reverse osmosis, nano and ultrafiltration systems, Proceedings of the American Waterworks Association (1997) Membrane Technology Conference, February 23-26, New Orleans, U.S.A 979-999.

3.   Ś.F.E. Boerlage, M.D. Kennedy, P.A.C. Bonne, G. Galjaardand J.C. Schippers, Prediction of flux decline in membrane systems due to particulate fouling, Desal., (1997), 113, 231-233.

4.   Ś.F.E. Boerlage, M.D. Kennedy, M.p. Aniye, E.M. Abogrean, G.Galjaard and J.C. Schippers, Monitoring particulate fouling in membrane systems, Desal., 118, (1998),131-142.

5.   Ś.F.E. Boerlage, M.D. Kennedy, G.J. Witkamp, J.P. Van der Hoek and J.C. Schippers, BaSO$_4$ solubility in reverse osmosis concentrates, J. Mem. Sci., 159 (1999), 47-59.

6.   Ś.F.E. Boerlage, M.D. Kennedy, G.J. Witkamp, I. Bremere, J.P. Van der Hoek and J.C. Schippers, Stable Barium Sulphate Supersaturation in Reverse Osmosis, J. Mem. Sci., 179, (2000) 53-68.

7.   Ś.F.E. Boerlage, M.D Kennedy, M.p. Aniye, E.M. Abogrean, D.Y. El-Hodali, Z. S. Tarawneh and J.C. Schippers Modified Fouling Index $_{-Ultrafiltration}$ to Compare Pretreatment Processes of Reverse Osmosis Feedwater, Desal., 131, (2000), 201-214.

8.   Ś.F.E. Boerlage, M.D. Kennedy, G.J. Witkamp, J.P. Van der Hoek and J.C. Schippers, Scaling prediction in reverse osmosis systems: Barium sulphate. Submitted to: J. Mem. Sci. (2000).

9.   Ś.F.E. Boerlage, M.D. Kennedy, M.R. Dickson and J.C. Schippers, The Modified Fouling Index using ultrafiltration membranes (MFI-UF): characterisation, filtration mechanisms and membrane selection, Submitted to: J. Mem. Sci. (2001).

10. Ś.F.E. Boerlage, M.D Kennedy, M.p. Aniye, Z. S. Tarawneh and J.C. Schippers, The MFI-UF as a Water Quality Test and Monitor, In preparation.

11. Ś.F.E. Boerlage, M.D Kennedy, M.p. Aniye, Z. S. Tarawneh and J.C. Schippers, Applications of the Modified Fouling Index$_{-Ultrafiltration}$, In preparation.

Curriculum vitae

Siobhán Francesca Emmerentianna Boerlage was born in Auckland, New Zealand (Aotearoa). She received her Bachelor of Science in Genetic Engineering in 1987 from the University of Auckland and continued studying at the University of Auckland within the Bachelor of Arts and Bachelor of (Civil) Engineering programmes. In 1988 Siobhán joined Beca Carter Hollings and Ferner Consultancy as an environmental scientist and studied part time at the University of Auckland, completing papers from the Masters of (Civil) Engineering programme. Leaving New Zealand in 1990, she began working at Birkbeck College the University College of London in England as an MSc Lab co-ordinator. In 1993 she was conferred the degree of Master of Science with distinction in Advanced Environmental Sanitation at the International Institute for Infrastructural, Hydraulic and Environmental Engineering (IHE) in Delft, The Netherlands. For the latter degree she wrote her thesis on the "Chemisorption of Dissolved Oxygen onto Granular Activated Carbon" within the field of drinking water treatment. In 1995 Siobhán began her PhD research which is presented here.